NATIONAL DEFENSE RESEARCH INSTITUTE

T0146371

An Assessment of Fiscal Year 2013 Beyond Yellow Ribbon Programs

Laura Werber, Jennie W. Wenger, Agnes Gereben Schaefer, Lindsay Daugherty, Mollie Rudnick

Prepared for the Office of the Secretary of Defense

For more information on this publication, visit www.rand.org/t/rr965

Library of Congress Control Number: 2015939209

ISBN: 978-0-8330-8896-3

Support RAND

Make a tax-deductible charitable contribution at
www.rand.org/giving/contribute

www.rand.org

Preface

In fiscal year (FY) 2011, Congress appropriated $16 million to expand Yellow Ribbon Reintegration Program outreach and reintegration services across the country—this supplemental appropriation has been termed *Beyond Yellow Ribbon* (BYR) funding. Initially, eight states were provided BYR supplemental funding, and in FY12, that number expanded to 12 states and one territory. In FY13, ten states received the funding: California, Colorado, Florida, Indiana, New Hampshire, North Carolina, Oregon, Tennessee, Vermont, and Washington. These funds were distributed to individual states, and the programs tend to be organized within the state National Guard structures.

These Beyond Yellow Ribbon programs provide a variety of support resources to service members returning from deployment, including employment counseling, behavioral health counseling, suicide prevention, and referrals to other providers, among other services. States had considerable latitude to design programs based on the specific issues and challenges faced by service members in their states.

In response to a congressional request to identify programs with strong records of success and to develop a nationwide set of promising practices, in October 2013, the Office of the Assistant Secretary of Defense for Reserve Affairs asked the RAND Corporation to carry out a formal review of 13 programs in the ten states that received FY13 BYR supplemental appropriations. The 13 programs included in this review are

- California: Work for Warriors
- Colorado: Marketing and Outreach Program
- Florida: Family Career Connection
- Indiana: Employment Coordination Program
- New Hampshire: Deployment Cycle Support Program
- North Carolina: Integrated Behavioral Health System, Education and Employment Center, Legal Assistance
- Oregon: Joint Transition Assistance Program, Military Assistance Helpline
- Tennessee: Employment Enhancement Program
- Vermont: Veterans Outreach Program
- Washington: Employment Enhancement Program.

RAND used a case study approach to document each program's resources, activities, and outputs; assess whether the program met its stated goals; and identify promising practices and areas for improvement.

This research should be of interest to federal and state policymakers, resource providers, other states that may have similar programs in place, and others concerned with how to improve support for service members.

This research was sponsored by the Office of the Assistant Secretary of Defense for Reserve Affairs and conducted within the Forces and Resources Policy Center of RAND's National Defense Research Institute, a federally funded research and development center sponsored by the Office of the Secretary of Defense, the Joint Staff, the Unified Combatant Commands, the Department of the Navy, the Marine Corps, the defense agencies, and the defense Intelligence Community. For more information on the RAND Forces and Resources Policy Center, see http://www.rand.org/nsrd/ndri/centers/frp.html or contact the director (contact information is provided on the web page).

Questions and comments regarding this research are welcome and should be directed to the leaders of the research team: Laura Werber (Laura_Werber@rand.org), Jennie Wenger (Jennie_Wenger@rand.org), or Agnes Gereben Schaefer (Agnes_Schaefer@rand.org).

Contents

Figures

Summary

Programs funded by Beyond Yellow Ribbon (BYR) began as an expansion of the Yellow Ribbon Reintegration Program (YRRP), a congressionally mandated U.S. Department of Defense (DoD) effort established in 2008 to provide deployment cycle information, services, and referrals to reserve-component personnel and their families throughout the deployment cycle. In 2011, Congress appropriated funding to expand the YRRP by authorizing service- and state-based programs to provide access to service members and their families of all components. This supplemental appropriation served to fund BYR programs, which are intended to provide critical outreach services to personnel returning from deployments and to their families. BYR's overall goal is to ease service members' transition back into civilian life. In response to a congressional request to identify programs with strong records of success and to develop a nationwide set of promising practices, the Office of the Assistant Secretary of Defense for Reserve Affairs asked the RAND Corporation to provide an assessment of 13 programs in the ten states that were supported by fiscal year (FY) 2013 BYR funding. The objectives of our study were to (1) examine the extent to which BYR programs have met their stated goals and the degree to which they have been effective in supporting reserve-component service members and their families, (2) identify promising practices in the programs that could be transferred across the broader set of BYR programs, and (3) suggest ways to improve the effectiveness of those programs as a whole.

Background: Beyond Yellow Ribbon Programs

The 13 programs supported by FY13 BYR funding vary widely in their scale, scope, focus, and services provided. Figure S.1 lists the name and location of each program in our study and the level of BYR funding it received in FY13. The focus areas of the 13 programs are as follows (some programs focus on multiple areas):

- employment (7 programs)
- behavioral health issues (2)

Figure S.1

FY13 Beyond Yellow Ribbon Programs and Appropriations

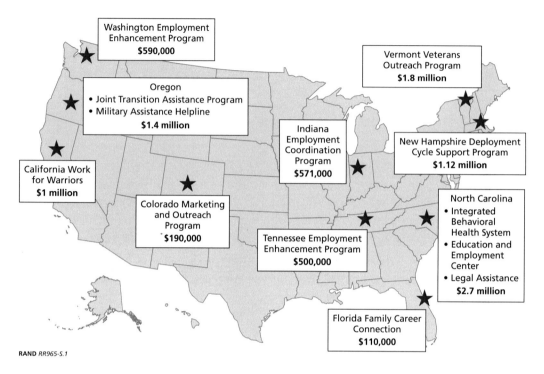

RAND RR965-S.1

- marketing and outreach (2)
- holistic, providing wraparound support (2)
- legal assistance (1).

All programs have an inclusive approach toward eligibility, offering support to members of the National Guard, Reserves, Active Component, and veterans from all conflicts. Some programs also support spouses and other family members. However, we found that programs tend to target guard personnel or veterans who have completed their military service. Prospective clients are made aware of programs through a variety of outreach strategies, including websites, social media, print media, and word of mouth. In addition, outreach occurs at events (e.g., YRRP events, job fairs), at armories, within units, and when personnel are still deployed.

Approach

The programs included in our study were selected by our research sponsor and represent the full set of states that received BYR funding in FY13. For our assessment of these programs, we used a standardized case study approach. Our assessment was based on the following activities, with a one-day site visit to each program headquarters at the heart of our approach:

- pre–site visit interview with the program point of contact
- review of available program documentation, such as a program logic model or conceptual framework, marketing materials for prospective clients, records of program outputs or outcomes, and program evaluation or audit reports
- on-site interviews with program staff
- on-site feedback session with program staff.

The site visits were coordinated through the efforts of the National Guard Bureau and took place January through April 2014.

Through interviews before and during the site visit, we collected information on the following topics:

- program background: such details as how long the program has been in place and how it is administered
- goals: what the program aims to accomplish
- inputs: resources, such as people, money, and materials
- activities: what the program does to reach and help guard and reserve personnel, veterans, and other clients
- outcomes: how clients are affected by the program
- impacts: how the state, DoD, and society are affected by the program
- facilitators and challenges: what has helped the program to meet its goals, and what has made it harder to do so
- promising practices: practices that have worked well for the program and may be useful to other programs.

After each site visit, we used our interview notes and the materials provided by program staff to prepare detailed program write-ups. These write-ups served as the basis for our individual program assessments, and we also use them for the cross-program analysis. Specifically, we coded the write-ups (that is, tagged specific passages of text) to identify cross-cutting themes related to facilitators, challenges, promising practices, and areas for improvement. Below we include a brief summary of each state's program(s) and our assessment of the extent to which each program is meeting its stated goals. Many of the programs have multiple goals. In cases of programs meeting some goals but not yet meeting others, we describe the goals as "partially met."

RAND's State-by-State Assessment

California

California Work for Warriors is a program that seeks to reduce unemployment by placing service members in jobs. The program's model depends on developing strong relationships with businesses that are willing to hire service members, identifying open job positions, and assisting service members throughout the application process. Work for Warriors gauges its success using job placements, but it also examines many other measures, such as the number of résumés on file, the number of business partnerships, and client satisfaction. The program uses these data to drive decisionmaking and program improvement. RAND found that the program is largely meeting its stated goal of placing as many individuals as possible, although the evidence is not sufficient to determine whether the initial goal of reducing unemployment among California National Guard personnel is being met. The program has a strong alignment among mission, activities, and measured outcomes that is likely helping to drive program success.

Colorado

The Colorado Marketing and Outreach Program is among the newest of the BYR-funded programs; at the time of RAND's visit, the program had been operating for about four months, and operating in steady state (stable condition) for only about one month. Primary goals of this program are to assist the Colorado National Guard (CONG) family programs in reaching their target populations and to increase utilization of those programs. Main activities include revamping the CONG Family Program website, establishing the CONG Family Program social media presence, and providing day-to-day marketing support for the other Family Program offices. Program staff also devote their time to developing medium- and long-term plans, procedures, and needs assessments. The program is admirable for its use of a wide spectrum of Internet and social media options to market CONG family programs and reach out to new clients. Given the short time in operation, RAND concluded that this program had not yet met its goals. Moreover, it is not entirely clear how the program will ultimately demonstrate that it has met its stated goals; the program lacks meaningful measures related to outcomes and impacts, as well as processes to collect these measures.

Florida

The Florida Guard Family Career Connection program focuses on a range of employment-related activities, and the program devotes considerable resources to networking with businesses. While this program has set a single clear goal of reducing guard unemployment, RAND's assessment is that data collected on statewide National Guard unemployment appear to be insufficient to determine whether the goal has been met. However, the program demonstrates strong relationships with businesses and focuses on those businesses that are committed to hiring service members. The personal relationships with units, service members, and employers help to ensure trust, usage, and referrals in this program.

Indiana

The Indiana Employment Coordination Program was designed as a one-stop shop for employment, but the program's goals also include supporting service members and reducing both unemployment and suicide rates in the National Guard. In RAND's assessment, this program is partially meeting its goals. In particular, the program is achieving placements, yet some of the program's other goals are not well defined and lack clear benchmarks. However, the program demonstrates several promising practices, including hosting relatively small employer events to improve job placement success rates.

New Hampshire

Emphasizing suicide prevention, mental health, employment, and homelessness, the New Hampshire Care Coordination Program aims to identify unmet needs of service members, family members, and veterans and provide the necessary support to meet those needs. The program works to accomplish this goal by referring clients to relevant agencies, but also by providing direct support. The program appears to have met the goals it measures related to suicide prevention, access to mental health care, employment, and homelessness. Furthermore, New Hampshire's program employs some promising practices, including hiring staff who have preexisting relationships with social service organizations and maintaining an extensive network of resource providers.

North Carolina

North Carolina has three BYR-funded programs. The goal of the Integrated Behavioral Health System (IBHS) program is to assess service members for immediate behavioral health needs, offer therapeutic support, provide case management services, and provide referrals to federal, state, and local resources. In RAND's assessment, IBHS is meeting its stated goals. Also, this program's robust data collection system and routine use of data are promising practices with implications for other programs. The Education and Employment Center works to establish relationships with veteran-friendly employers in North Carolina and to place service members in civilian jobs that help achieve their career goals. RAND's assessment is that this program is partially meeting its goals; however, this is primarily because the program is so new. In particular, the program is meeting its short-term goals of establishing relationships with employers and placing service members in civilian jobs. Finally, the North Carolina National Guard Legal Assistance program works to provide service members with low- or no-cost access to legal services. RAND's assessment is that Legal Assistance is meeting its stated goal. The single point of entry utilized across these programs is a promising practice.

Oregon

Oregon has two BYR-funded programs—the Joint Transition Assistance Program (JTAP) and the Oregon Military Assistance Helpline (hereafter, the Helpline). JTAP provides a wide range of employment services, with a short-term goal of achieving 800 job placements in 2014 and a long-term goal of improving Oregon National Guard readiness and retention rates. The Helpline's goal is to provide 24-hour access to counseling services and to prevent service member suicides. In RAND's determination, the Helpline has met its stated goal of providing around-the-clock counseling, and JTAP is partially meeting its stated goals and was on track to meet its 2014 placement goal. However, the impacts of these programs' activities on readiness or retention are less clear. JTAP's placement within the Military and Family Readiness Directorate (J9) and the integration of all J9 programs facilitates effective delivery and constitutes a promising practice.

Tennessee

The Tennessee National Guard Employment Enhancement Program (NGEEP) is among the newest of the BYR-funded programs; at the time of RAND's visit, the program had been operating for about four months. The primary goals of NGEEP include improving employment outcomes for service members and their families, as well as enhancing readiness and resilience. Along with providing many activities in support of the employment goal, the program's job counselors also spend a significant amount of time working with employers in their assigned region of the state. In RAND's determination, the program's goals are partially met. The program has clearly placed service members in jobs, but it does not have the metrics in place to measure an effect on readiness or resilience—something that would be quite challenging given NGEEP's limited scope. Although the program is new, its model of hiring program staff to handle specific geographic regions and giving them discretion to tailor their programs based on local employment conditions holds promise.

Vermont

The Vermont Veterans Outreach Program (VTVOP) is one of the most established programs included in this study. Its goals are to connect Vermont's veterans to the support services that they need and to enroll veterans in the federal benefits for which they are eligible. VTVOP is a wraparound program that offers highly personalized support to veterans, which may include a combination of counseling, case management, and referrals, with house calls and transportation assistance as appropriate. Based on RAND's assessment, VTVOP is meeting its stated goals. The program's robust data collection systems and extensive training programs for its outreach specialists constitute promising practices.

Washington

Washington's Employment Enhancement Program (EEP) aims to produce employment-ready service members within 30 days, to change the way service members seek employment, and, ultimately, to place them in jobs. The EEP focuses on a variety of employment-related activities in a program that staff describe as being "high touch" rather than "high tech." In RAND's assessment, the program has met at least some of its goals: Hundreds of clients have found employment after receiving support from EEP staff. The program's limited tracking of outputs and outcomes makes it difficult to assess whether the program has met all of its stated goals. The program's emphasis on high-touch support and its focus on empowering clients by improving their employment search skills are promising practices that may be transferrable to other states.

RAND Observations Across Programs

In RAND's determination, nearly all of the BYR programs are at least partially meeting their goals. Despite the diversity among the programs, we uncovered numerous common themes during our site visits and interviews with program staff. We group these themes into facilitators, challenges, promising practices, and areas for improvement. We regard facilitators and challenges as factors generally beyond the programs' control but with the potential to influence the programs' abilities to meet their goals. Promising practices and areas for improvement also are likely to influence programs' abilities to meet their goals, but these aspects of program operations generally are within the programs' purview.

Common examples of each include the following:

- Facilitators
  - *Strong leadership support.* Support from political and military leaders appears to be a facilitator for several of the programs. In contrast, some programs describe a situation with limited or insufficient support, which constitutes a challenge.
  - *Physical and organizational integration.* This also serves to facilitate success. In particular, several programs are organized as part of the J9 directorate, and some programs are physically located near other family programs.
  - *Access to free or low-cost technological resources.* Several programs report that the ability to utilize low-cost or free technological resources facilitates meeting their goals. An example of such a resource is employment databases. Some programs also utilize resources provided by the command, such as office space.
- Challenges
  - *Insufficient leadership support.* In several states, the backing from leaders at different levels has been inconsistent, rendering program outreach and data collection efforts more difficult.

- *Program overlap.* Some programs overlap with programs supported by other funding sources. This challenge seems to occur mostly in the earliest phases, and several programs report experiencing but successfully overcoming this challenge.
- *Service members' unwillingness to fully utilize support services.* This theme emerges especially among those programs focused on employment; staff of such programs mentioned motivation as a challenge.
- *Funding.* The amount, timing, uncertainty, and logistics of funding pose a challenge for BYR programs; funding was the most prominent challenge reported.
- *Staff turnover.* This is a challenge somewhat related to the one above. Many programs are small in scale, and staff often possess specialized knowledge or close working relationships with employers, social service agencies, or other resource providers. Funding uncertainties, as well as job-related stress, have the potential to create high levels of turnover, which would pose a significant challenge to these programs.

- Promising practices
  - *Close alignment between activities and goals.* Alignment promotes the efficient use of resources to achieve program outcomes.
  - *Single point of entry.* Funneling service members into a program via one entry point helps to link them to services seamlessly and facilitates tracking them as they receive services.
  - *High-touch, one-on-one services.* One-on-one services can distinguish a program from other support options and may help some service members beyond what can be accomplished by technological resources alone.
  - *Geographically dispersed staff.* Having staff spread throughout the state facilitates connections with service members and with other community partners; it also encourages program staff to be more aware of region-specific factors.
  - *Activities that build long-term client skills.* By doing more than just meeting the immediate needs of service members and other clients, programs can help them to avoid future challenges or tackle them independently.
  - *Strong partnerships with other resource providers.* These partnerships can facilitate referrals and more-comprehensive support when needed, and they have the added benefit of helping to avoid resource redundancy.
  - *Strong partnerships with community organizations and employers.* These partnerships can be force multipliers, especially when resources are limited.
  - *Early engagement with mobilized units.* Many programs reported successful outcomes when they first connected with units returning from deployment while still in theater or within weeks of their return.
  - *Broad and creative outreach strategies.* Making use of free or low-cost technology and social media like Facebook, Linked In, and Twitter helps reach and engage more service members.

- *Strong technology-based tracking systems.* These systems are useful for client case management and capturing outputs and outcomes.
- *Efforts to make programs replicable and sustainable.* Creating documentation on program activities, procedures, and processes ensures that programs can be sustained regardless of changes in staffing. Also, prioritizing and phasing contracts could help sustain these programs in times of uncertain funding.
- Areas for improvement
  - *Lack well-defined, measurable goals.* A program should define clear, measurable goals, and these goals should drive the program's activities and how it demonstrates results in both the short and long term.
  - *Insufficient evidence of outcomes and impacts.* Programs whose goals are not clearly defined or are difficult to measure will lack evidence of influence, and even the programs with well-defined, measurable goals and strong data collection systems often lack subjective data to round out their analyses.
  - *Insufficient outreach to the entire eligible population.* Programs generally reported that they do not turn service members away, but there often is a lack of outreach to the entire eligible population. We recognize the inherent tension between providing one-on-one services and serving the entire population; in some cases, programs may need additional resources or may need to alter their services as a result of increased outreach.
  - *No contingency plan for the limitations of BYR funding.* While all of the programs report that the uncertainty of BYR funding constitutes a challenge, in some cases, programs did not fully consider the limitations of such funding. For example, some programs spent a substantial portion of the year developing (rather than executing) programs; many use the BYR funds to support critical programs but do not search for alternative funding to guard against uncertainty.

Recommendations

We divide our recommendations into two sets: (1) those that can be instituted at the program level and (2) those that may be helpful in assisting DoD or congressional policymakers as they consider general program oversight and future BYR funding allocations.

Recommendations for Program Leaders

We have the following recommendations for program leaders:

- Develop meaningful, measurable goals.
- Collect and learn from program data on effectiveness.
- Ensure that programs are sustainable.
- Utilize practices associated with high-quality programs.

First, we recommend that programs develop meaningful, measurable goals based on their specific activities and outcomes. Programs should seek to measure what they *do* (outputs) and what they *get* (outcomes). The programs we reviewed offer some examples of concrete output metrics, such as website hits, presentations given, activities attended, telephone calls placed, résumé/interview skill sessions, referrals, counseling sessions per staff member, job applications submitted, and tax returns filed. The greater challenge is developing concrete outcome measures. We recommend that programs collect measures that provide a robust picture of outcomes, both short term and longer term. The programs we reviewed also offered some examples of concrete outcome measures that other programs could emulate. For instance, some employment programs measure job placements, average hourly wage, cost per placement, client satisfaction, and improved job search skills (including networking, résumé development, and interviewing). We found that programs focused on areas other than employment also measure concrete outcomes, such as the number of interventions conducted to stabilize a crisis situation or otherwise assist service members and their families, number of clients receiving mental health treatment, and increased program utilization. Goals also need to be realistic, given both capacity and time frame (due to annual BYR funding uncertainty).

Second, we recommend that programs collect and learn from program data on effectiveness. Collecting meaningful data will allow programs to determine whether their goals are being met. Thus, data collection should flow directly from the program's goals. For instance, some programs only conduct activities that directly align with their goals, and then, in an effort to incentivize focus on these key activities, program leaders track outputs and outcomes that they believe are closely tied to their goals. Collecting and reviewing such data should allow programs to pick up on trends and to make adjustments to resources as necessary. We found that some programs have been analyzing the data that they collect, which has enabled them to be quite successful at refining their activities and metrics over time to more closely align with their primary goals. Collecting such data, however, is not sufficient to determine causality; assessing whether these programs actually *bring about* desired changes requires a formal evaluation involving a control group (a group of people who are very similar to program participants but who do not receive the program's services). A final point is that some programs face specific barriers to data collection. For example, they may lack the authority to collect the appropriate data, they may lack the time or resources to collect the data, or they may struggle with formatting or storage issues (such as data that exist only on paper or in difficult-to-aggregate formats). Program leadership or state-level interventions may be necessary to resolve these problems.

Third, we recommend that programs put a greater focus on ensuring program sustainability. The uncertain nature of BYR funding poses a substantial challenge for all of the programs; a loss or delay of funding would likely mean at least a temporary closure of many programs and thus the loss of program services. While program directors have only limited ability to control funding or turnover, there are practices that can serve to make programs more sustainable even in light of such uncertainty. We found that a few states have made concerted efforts to ensure the sustainability of their programs by carefully documenting the programs' activities, including expectations for standard operating procedures (SOPs), effective or promising practices, and challenges the program has faced, as well as solutions identified to address these challenges. In addition, programs may be able to establish alternative funding streams or sources; these could serve as buffers against funding uncertainty. Some programs also indicate that their success relies largely on individual personalities. The practice of creating and utilizing SOPs can help to retain knowledge even in the face of staff turnover. Programs will also likely benefit in this regard from clearly defined staff roles, and some programs may benefit from providing additional staff training. Finally, encouraging or requiring that program staff keep detailed records of contacts with service members and other stakeholders may help to improve program sustainability from a case management standpoint, even in the face of staff turnover.

Finally, we recommend that program leadership work to adopt and utilize practices associated with high-quality programs. In the literature, high-quality programs are described as ones that are evaluated, sustained, and replicable. Accordingly, the aforementioned use of SOPs is associated with high-quality programs because the documentation helps to ensure that the program will persist in light of staff turnover and that the same successful practices will be used consistently. Efforts to ensure program viability, even in light of BYR funding decreases, are also a high-quality program practice, one focused on sustainability. Another practice associated with high-quality programs is ensuring that services provided to clients are evidence-based. Finally, the use of an evaluation process (as described above) is linked to improved decisionmaking, oversight, and monitoring.

Recommendations for Department of Defense and Congressional Policymakers

We have the following recommendations for DoD and congressional policymakers:

- Address programs' concerns regarding BYR funding.
- Clarify appropriate use of BYR funds.
- Share promising practices across programs.
- Encourage programs to widen their focus beyond the National Guard.

First, we recommend that leadership work to address programs' concerns regarding the uncertainty of BYR funding. We recognize that DoD leadership is well aware of this issue and that to some extent, this is a function of the current budget environment. However, we recommend leadership do whatever possible to transfer funds in a timely manner and to encourage programs to develop contingency plans. In the same vein, leadership could provide feedback to programs to help them determine the most appropriate use of funds. For example, these funds may be especially appropriate for purchasing equipment or carrying out other one-time investments to provide services. Alternatively, these funds might work well to pilot test innovative ideas, especially if there are other potential sources for proven programs.

Second, we recommend that leadership clarify the appropriate use of BYR funds. The uncertainty of BYR funding also suggests that thinking strategically about the types of programs supported by BYR funds, and, more generally, about how BYR funds should be used, may be especially appropriate. Some programs funded with BYR dollars may be viewed as critical, while others may be considered add-on. For a critical program, the end of services or even a lapse in service could be extremely problematic. Along the same lines, other uses of these funds should be considered. For example, the funds may work well for pilot programs that intend to find other sources of funding if they prove successful. Ultimately, DoD and congressional policymakers should consider providing guidance to the states on more—and less—appropriate uses of BYR funding.

Third, we recommend that leadership work to share promising practices across the BYR programs. We found that programs are very eager to learn what other programs are doing. Therefore, sharing information on promising practices (as well as on barriers and the extent to which different programs have overcome barriers) is likely to be of interest to all programs and could be especially helpful for new programs. We encourage DoD and congressional policymakers to share this information in a variety of ways; examples include encouraging cross-program communication and collaboration (perhaps via web-based meetings or an in-person meeting for program staff and leaders). Sharing barriers, even if programs have not found ways to overcome these issues, may be useful as well.

Finally, we recommend that programs be encouraged to expand their focus beyond the National Guard. Some programs focus primarily on National Guard, or even Army National Guard, personnel in their states. We recommend that leadership encourage programs to widen their focus and their marketing efforts to include more service members, veterans, or family members, as appropriate. If it is the intention that these programs serve both the National Guard and Reserves, this should be communicated to programs that are currently focused primarily on the Guard, and efforts should be made by the programs to expand services.

Acknowledgments

We thank the office of the Assistant Secretary of Defense for Reserve Affairs for its support. Specifically, we thank our project monitor, Peter Weeks, who worked closely with us throughout the course of the project, and Marie Balocki, of the Yellow Ribbon Reintegration Program, who provided helpful guidance and suggestions as we conducted our research.

We greatly appreciate the efforts of the National Guard Bureau, specifically BG Marianne Watson (director, Manpower and Personnel, National Guard Bureau). Brigadier General Watson introduced the study to the adjutants general in all the states within the scope of this research effort, thereby facilitating our site visits to all the programs.

Thomas L. Bush, Gabriella Gonzalez, and Tepring Piquado provided formal peer reviews that ensured our work met RAND's high standards for quality assurance. We also benefited from the contributions of RAND colleagues. Shaela Moen, Robert Stewart, and Jonathan Wong contributed to the site visit analysis. Michelle McMullen and Donna White provided administrative support.

Finally, we note that we could not have completed this work without the assistance of the leadership for each of the programs included in our assessment and the participation of program staff in confidential interviews.

We thank them all, but we retain full responsibility for the objectivity, accuracy, and analytic integrity of the work presented here.

Abbreviations

AAR	After-Action Review
BYR	Beyond Yellow Ribbon
CASY	Corporate America Supports You
CCP	Care Coordination Program
CONG	Colorado National Guard
CPRS	Computerized Patient Record System
DoD	Department of Defense
DPH	director of psychological health
EEC	Education and Employment Center
ECP	Employment Coordination Program
EEP	Employment Enhancement Program
ESGR	Employer Support of the Guard and Reserve
ETC	employment transition coach
FAC	family assistance coordinator
FGFCC	Florida Guard Family Career Connection
FY	fiscal year
H2H	Hero2Hired
IBHS	Integrated Behavioral Health System
IDVA	Indiana Department of Veterans Affairs

INNG	Indiana National Guard
J1	Manpower and Personnel Directorate
J9	Military and Family Readiness Directorate
JBLM	Joint Base Lewis-McChord
JCEP	Job Connection Education Program
JSS	Joint Services Support
JSSWA	Joint Services Support Washington
JTAP	Joint Transition Assistance Program
LVER	Location Veterans Employment Representative
MSCCN	Military Spouse Corporate Career Network
NCNG	North Carolina National Guard
NCO	noncommissioned officer
NGEEP	National Guard Employment Enhancement Program
NGEN	National Guard Employment Network
NHDHHS	New Hampshire Department of Health and Human Services
NHNG	New Hampshire National Guard
O*NET	Occupational Information Network
ORNG	Oregon National Guard
SMFS	Service Member and Family Services
SOP	standard operating procedure
SRP	Soldier Readiness Processing
TAA	transition assistance advisor
USERRA	Uniformed Services Employment and Reemployment Rights Act
VA	Department of Veterans Affairs
VistA	Veterans Health Information Systems and Technology Architecture

VTANG   Vermont Air National Guard

VTARNG   Vermont Army National Guard

VTNG   Vermont National Guard

VTVOP   Vermont Veterans Outreach Program

WANG   Washington National Guard

WOWI   World of Work Inventory

YRRP   Yellow Ribbon Reintegration Program

CHAPTER ONE

Introduction

The Yellow Ribbon Reintegration Program (YRRP) is a Department of Defense (DoD) initiative established in 2008 to provide deployment cycle information, resources, programs, services, and referrals to reserve-component personnel and their families. YRRP offers a series of events for personnel and their families that occur throughout the deployment cycle, from predeployment, during deployment, and 30, 60, and 90 days postdeployment. These events provide information and services to support guard and reserve personnel and their families. The goals of the program are to maximize successful transitions as personnel move between their military and civilian roles and to create strong, resilient military families.[1]

In fiscal year (FY) 2011, Congress appropriated $16 million to expand YRRP outreach and reintegration services across the country, and this supplemental appropriation has been termed *Beyond Yellow Ribbon* (BYR) funding. Initially, eight states were provided BYR supplemental funding (Colorado, Minnesota, New Hampshire, New Jersey, North Carolina, Oregon, Vermont, and Washington), and in FY12 that number expanded to 12 states and one territory (adding Florida, Guam, Nevada, West Virginia, and Wyoming). In FY13, ten states received BYR funding (California, Colorado, Florida, Indiana, New Hampshire, North Carolina, Oregon, Tennessee, Vermont, and Washington). States receiving BYR funding use this resource to provide a variety of support to personnel returning from deployment, including those related to employment, behavioral health counseling, and suicide prevention.

In March 2012, DoD solicited metrics to help assess the effectiveness of BYR-funded programs. A total of 12 services (e.g., employment, suicide prevention, financial support) were listed in the program assessment forms that each program was asked to complete. DoD found that most states receiving $1 million or more offered between ten and 12 services, and DoD identified five states that demonstrated "best practices" (Nevada, North Carolina, Oregon, Vermont, and Washington).[2]

---

[1] YRRP, *Annual Advisory Board Report to Congress, Fiscal Year 2011*, Washington, D.C., March 2012a, p. 2.

[2] YRRP, Supplemental Funding Update briefing, August 20, 2012b.

Study Purpose

In response to a congressional request to identify programs with strong records of success and to develop a nationwide set of promising practices, in October 2013, the director of YRRP in the Office of the Assistant Secretary of Defense for Reserve Affairs asked the RAND Corporation to carry out a formal assessment of 13 programs in the ten states that received FY13 BYR supplemental appropriations. Table 1.1 lists the programs that are included in our study and the total BYR appropriations that each state received in FY13.

The objectives of our study were to (1) examine the extent to which BYR programs have met their stated goals and the degree to which they have been effective in supporting reserve-component service members and their families, (2) identify promising practices in the programs that could be transferred to the broader set of BYR programs, and (3) suggest ways to improve the effectiveness of those programs. In order to achieve these objectives, we carried out a detailed assessment of each of the programs in our study. The following section provides an overview of our study approach.

Figure 1.1
FY13 Beyond Yellow Ribbon Programs and Appropriations

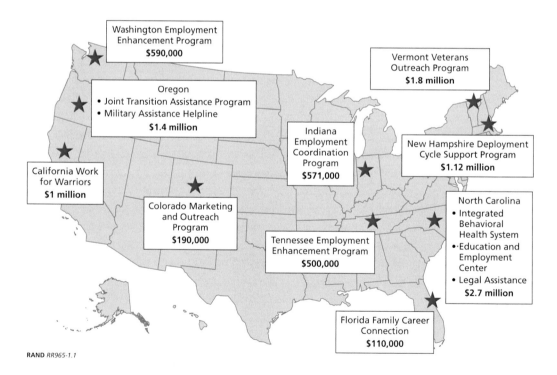

RAND RR965-1.1

Study Approach

The programs included in our study were selected by our research sponsor and represent the full set of states that received BYR funding in 2013. For our assessment of these programs, we opted to use a standardized case study approach. Although the generalizability of our results is limited, this research design enabled us to accommodate for differences in individual program scope and focus, while still providing the basis for a systematic, rigorous program assessment. We collected similar information about each program, integrated different data sources to assess the program, and then looked across the programs for cross-cutting themes, such as common facilitators and areas for improvement. Each program assessment consisted of the following activities, with a one-day site visit to each program headquarters central to our approach:

- pre–site visit interview with the program point of contact
- review of available program documentation, such as a program logic model or conceptual framework, marketing materials for prospective clients, records of program outputs or outcomes, and program evaluation or audit reports
- on-site interviews with program staff
- on-site feedback session with program staff.

The site visits were coordinated through the efforts of the National Guard Bureau. Specifically, BG Marianne Watson (director, Manpower and Personnel, National Guard Bureau) sent a letter to the adjutants general in all the states included in our study. In her letter, Brigadier General Watson explained the purpose of the study and requested that the states provide RAND with a point of contact with whom the RAND study team could coordinate its site visit and data collection requests. The ten site visits took place January through April 2014.

To develop our interview protocols for this project, we consulted the literature related to program review, program evaluation, and logic models.[3] The protocols covered the following topics:

- program background: details such as how long the program has been in place and how it is administered
- goals: what the program aims to accomplish
- inputs: resources, such as people, money, and materials

---

[3] The works we relied on the most were Joie Acosta, Gabriella C. Gonzalez, Emily M. Gillen, Jeffrey Garnett, Carrie M. Farmer, and Robin M. Weinick, *The Development and Application of the RAND Program Classification Tool, The RAND Toolkit, Volume 1*, Santa Monica, Calif.: RAND Corporation, RR-487/1-OSD, 2014; L. W. Knowlton and C. C. Phillips, *The Logic Model Guidebook: Better Strategies for Great Results*, Thousand Oaks, Calif.: Sage Publications, 2013; and Peter H. Rossi, Mark W. Lipsey, and Howard E. Freeman, *Evaluation: A Systematic Approach*, 6th ed., Thousand Oaks, Calif.: Sage Publications, 1998.

- activities: what the program does to reach and help guard and reserve personnel, veterans, and other clients
- outcomes: how clients are affected by the program
- impacts: how the state, DoD, and society are affected by the program
- facilitators and challenges: what has helped the program to meet its goals, and what has made it harder to do so
- promising practices: practices that have worked well for the program and may be useful to others.

A detailed description of our methods, including our interview questions and our program synthesis guide, is provided in the appendix to this report. Two RAND researchers went on each site visit, which was typically at the state's National Guard Joint Force Headquarters and entailed meeting with a program's director and its staff. We structured the site visits to include an initial interview session covering the program's background, goals, inputs, activities, and results. Following that, we took a break to rapidly prepare a set of slides that summarized what we learned during that session. Teams used a preestablished briefing template to develop the slides, which loosely followed the steps one would take to develop a program logic model, identifying goals, activities, and resources, as well as outputs and measures of success. When we reconvened for our second interview session, we presented the slides to the program director and staff, asking for corrections and using them to elicit more information about the program. Finally, in a typical visit, we wrapped up the site visit with questions about data collection and usage, program facilitators and barriers, and future plans. In some cases, the two sessions occurred on the same day; in other cases, the visit took up portions of two consecutive days.

Across programs, the number of people we met with ranged from two to nine. After each site visit, we used our interview notes and the materials provided by program staff to prepare detailed program write-ups. To ensure a consistent approach to these write-ups, we developed a *program synthesis guide* that included both detailed instructions about the program write-up process and a list of topics that should be covered in the write-up. The topics in the program synthesis guide corresponded to those featured in the interview protocols and covered background/orientation information for the program, areas for improvement, and promising practices. These individual write-ups served as the basis for Chapters Two through Eleven, in which we present our analyses of the programs in each of the ten states in our study.

The finalized program write-ups served as the basis for the cross-program analysis. Specifically, we coded the write-ups based on the topics in the program synthesis guide. Codes are used to tag passages in a data source so they can be easily retrieved later and organized by topic or other characteristics. After the coding was complete, we reviewed the results and drafted common themes and notable findings. For example, two researchers reviewed all the sections coded as "promising practices" and noted those common across several programs. The results of this cross-program analysis for facilitators, challenges, promising practices, and areas for improvement are featured in Chapter Twelve.

Organization of This Report

Chapters Two through Eleven present our assessment of the BYR-funded programs in California, Colorado, Florida, Indiana, New Hampshire, North Carolina, Oregon, Tennessee, Vermont, and Washington. Each chapter presents an overview of the program(s) that we reviewed in that state, including history, goals, target population, and resources (e.g., personnel, office space, and technology). Next, each chapter reviews program activities, outreach, and unique features. These chapters also review how the programs measure and evaluate their activities, including the types of information that they collect and how they use that information. Next, the chapters review the programs' results, including outputs, outcomes, and impacts, and they identify both specific challenges that each program has faced and facilitators that have helped the programs. Finally, each chapter presents RAND's assessment of each program, including whether the programs are meeting their stated goals, whether they have promising practices that may be helpful to other state programs, and any areas in which they could improve.

Chapter Twelve presents the findings from our analysis across all of the programs that we assessed. In this chapter, we summarize cross-cutting themes, including common facilitators and challenges. We also call attention to the practices that we believe both hold great promise and can be readily transferred to new and existing programs in other states, and we identify the most pressing areas for improvement across the programs.

Chapter Thirteen presents our conclusions and recommendations. In this chapter, we offer recommendations that could be implemented at the program level, as well as recommendations to DoD and congressional policymakers. The appendix includes details on our study approach, interview protocols, and program synthesis guide.

CHAPTER TWO

California Work for Warriors Program

Bottom Line Up Front

Work for Warriors is a program that seeks to place service members in jobs and reduce unemployment. These goals are addressed by developing strong relationships with businesses that are willing to hire service members, identifying open positions, and following service members throughout the application process. Work for Warriors gauges its success using a large and varied set of indicators, including outputs (such as applicants, résumés on file, and business partnerships) and outcomes (such as placements, cost per placement, and user satisfaction). The program is regularly using these data to drive decisionmaking and program improvement.

RAND found that the program is largely meeting its stated goals of placing as many individuals as possible in jobs, although the evidence is not sufficient to determine whether the initial goal of reducing unemployment between spring 2012 and spring 2013 was met.[1] The program appears to have a strong alignment among mission, activities, and measured outcomes that is likely helping to drive program success. The program's promising practices include its focus on a narrow mission, close partnerships with businesses, high-touch services (i.e., personal, intensive support of individual clients throughout the employment search process), and intentional use of metrics. Areas for improvement include needing greater coverage across the state and the Reserve Component, as well as needing to identify and attempt to measure some of the longer-term impacts of the program.

---

[1] Note that the Work for Warriors program did not receive BYR funds until after this period.

Overview

History and Goals

The Work for Warriors program was created in February 2012—a year prior to receiving federal funding from BYR appropriations.[2] The impetus for the program was the high level of unemployment among National Guard members returning from deployment, as well as a need to better connect service members to employment opportunities upon their return.[3] The initial goal of the program was to reduce unemployment in the California National Guard by 25 percent in one year.[4] The program is in the process of establishing goals that extend beyond the first year, and in the interim, program staff have been focused on achieving as many job placements as possible and minimizing cost per placement.

Target Population

At the time of our site visit, in January 2014, the program had largely focused on serving the approximately 22,500 members of the California National Guard, including both the Army National Guard and the Air National Guard, though program staff reported that they do not turn away service members or veterans who are not part of the National Guard. During our site visit, staff indicated that they have plans to expand and begin providing regular outreach and services to members of the Reserves.[5] Among guard personnel, there has been a particular focus on individuals who are soon returning or have recently returned from deployment. Program staff reported that they are constantly assessing the willingness of service members to work closely with them to achieve program goals, and they will focus efforts on individuals who demonstrate a commitment to finding employment. For example, if individuals are not responsive to efforts of program staff to provide assistance with résumés or interview skills after repeated attempts, or if they demonstrate repeated incidences of undesirable interview behavior (e.g., arriving late, being dressed inappropriately), the program will no longer make efforts to target services to these individuals.

Program staff also reported that the program's coverage of guard personnel across the state had been somewhat uneven, with more coverage in urban areas, particularly the north-central region of the state that includes the San Francisco Bay area and Sac-

---

[2]   Much of the information in this chapter is based on interviews with Work for Warriors staff conducted in January 2014, an informational presentation by the program director that same month, and additional materials that the staff provided. Specific sources are noted in subsequent footnotes.

[3]   Ty Shepard, director, California National Guard Employment Initiative, "Lowering the Rate of Unemployment for the National Guard and Reserve: Are We Making Progress?" testimony before a hearing of the House Committee on Veterans Affairs, March 14, 2013.

[4]   Shepard, 2013.

[5]   David S. Baldwin, Adjutant General, California National Guard, "Employment Initiative for California National Guard Members," memorandum, December 27, 2013.

ramento, and with some units making greater use of the program's services than others. The staff noted three reasons for these differences in coverage: (1) most of their open positions come from employers located in urban areas, and the program's services are centered on position openings; (2) until recently, the staff were only located in Sacramento, so it was easier to develop relationships with nearby businesses; and (3) the engagement levels of unit leadership vary, and some have been more active in referring service members to the program. The program staff reported that they are increasing efforts in Southern California, but are likely to continue focusing on urban areas, as this is where they are able to get the best results in terms of volume of job placements with a limited set of resources.

Resources

To establish and support the Work for Warriors program, in February 2012, the Speaker of the California Assembly provided $500,000 in funding to the California National Guard. State funding has persisted at the same level since the program's inception, and a bill was introduced in 2013 to establish funding for the program as a $575,000 line item in the state budget. In June 2013, the program received an additional $1 million in the form of a BYR congressional appropriation, and in January 2014, the program received word of renewed federal funding for the next fiscal year. While the program is largely funded through these state and federal allocations, it also receives donated office space and office resources, and occasional one-time donations from businesses to support particular efforts.

The program is positioned under the National Guard as a direct report to the adjutant general. Program staff report that the adjutant general was largely responsible for creating the program, and has been a strong supporter throughout. Initially, Work for Warriors was staffed by six individuals, including one program manager, two resource managers, one information technology specialist, one business coordinator, and an individual who was reassigned from state active duty to lead the effort. Federal funding enabled the program to add nine additional staff members, with most of the additions being resource managers who work directly with service members and businesses. According to program leadership, the important qualities they look for in Work for Warriors staff include prior job experience and an understanding of the networking process, an entrepreneurial spirit, internal motivation, and the ability to follow up. While staff are considered the program's "most important resource" and salaries account for the majority of program funds, some federal funding is also used to purchase technology to support the program.

Program staff reported that relationships with other organizations were also an important resource. The program relies heavily on its relationships with businesses to ensure a steady flow of open positions, and staff members devote substantial time to maintaining these relationships. In addition, staff cited frequent collaboration with the office focused on behavioral health (also positioned directly under the adjutant general), and a strong sense of support from all of the National Guard staff. Work for Warriors also strives to develop strong relationships with other nonprofits, programs serving veterans, and headhunting organizations to advertise the program and identify other means of meeting service member needs.

Finally, support from high-level leadership, including the adjutant general, the Speaker of the California State Assembly, and a Work for Warriors caucus in the U.S. House of Representatives, are all viewed as critical to maintaining recognition and funding for the program. In addition to ensuring that the program receives continued federal support, the bipartisan caucus led by Representative Mark Takano and Representative Paul Cook introduced a bill to expand the Work for Warriors program nationwide.[6]

Program Activities

Activities for Target Population

The primary focus of Work for Warriors is the direct placement of service members into jobs. The process is, first, to identify open positions through the relationships that program staff have developed with businesses. Then, staff widely advertise positions through social media and networks and identify individuals who are potentially interested in and qualified for the position through an initial screening process. Next, resource managers support individual service members through the full application process, including working with individuals to revise résumés and prepare for interviews, providing referrals and information on the candidates applying directly to the business, and following up with individuals and businesses (regardless of outcomes) to assess why the position was or was not filled by the service member. Communication with service members and businesses occurs through a combination of telephone, email, and social media.

With job placement as its primary focus, the Work for Warriors program devotes resources primarily to activities that it views as having a direct relationship with job placements. The importance of retaining this "singular focus" was mentioned numerous times by program staff during our interviews. For example, program leadership noted that the program no longer hosts or participates in job fairs, because job fairs are viewed as high-investment, low-return opportunities in terms of job placements. Rather than inviting businesses that may or may not have open positions and may or may not be committed to hiring service members, program staff instead focus on developing strong relationships with a limited number of businesses and identifying and advertising job opportunities as they emerge. In another example of the program's

---

[6] The bipartisan caucus is led by Representatives Takano and Cook. For press releases describing the caucus and bill, see Congressman Mark Takano, "Rep. Mark Takano and Rep. Paul Cook Form Bipartisan Work for Warriors Caucus," Riverside, Calif., November 8, 2013; and Congressman Mark Takano, "U.S. House of Representatives Passes Two Pieces of Legislation Submitted by Rep. Mark Takano," Riverside, Calif., May 22, 2014.

singular focus, the staff explained to us that they provide access to training opportunities only when they can see a direct connection to job placements. For example, many service members have expressed interest in positions in security-related occupations, so the program provides training for service members to receive their "Guard Cards," which are required for security guards in California.

The one activity that does not directly support job placements but was reported as an important component of the program is referring service members to other programs when they have needs that are not employment-related. Program staff noted that service members with employment issues often also experience mental health issues and are often in need of additional resources to support themselves or their families. These individuals are referred to the wide network of service providers with which the program has developed relationships.

Work for Warriors staff also described their "headhunter," or executive placement, model of achieving placements as an important feature of the program. Initially, the resource managers started with service members who were looking for employment and then made efforts to find jobs for which the individuals might apply. Now, they focus on building business relationships to find strong job placement leads and then matching this position to a strong set of service member candidates. The program focuses on developing partnerships with businesses that are truly committed to hiring service members. Program leadership describes the program as acting almost as an extension of the human resource departments for the businesses they partner with. Program leadership argued that by starting with strong job placement leads, Work for Warriors is able to be more efficient and has higher success rates because program staff are more certain that the business partner is committed to hiring a service member.

In fact, program staff often emphasized the need to support businesses as equivalent in importance to supporting service members; the program needs to maintain strong relationships with businesses and be seen as a trusted partner that has the interests of the business in mind as well. While some of the program's activities address service member needs, they are also intended to cultivate these business partnerships. For example, the program screens service members to ensure that the program is not sending businesses candidates who are unqualified or have demonstrated unprofessional behavior in past application efforts. In addition, the program conducts follow-up interviews with businesses both to get feedback on the service members who apply for positions and, in cases where a service member is not hired, to understand how the program can better meet the needs of the business.

Outreach

The program uses a range of efforts to provide outreach to service members. The first is a strong social media presence. Work for Warriors uses Facebook, LinkedIn, and Twitter to develop a community, advertise job openings and other relevant information, and communicate with service members. However, program staff report that telephone and email remain the primary means of communication. A staff member has been assigned responsibility for providing content for and monitoring social media, and the staff member reported a strong focus on ensuring that the outreach is having the maximum possible effect. To improve outreach efforts, the staff posts regularly to keep service members engaged and monitors metrics to identify the types and timing of posts that achieve the greatest levels of achievement (measured by "likes" and "shares" of posts).

In addition to social media, the program relies heavily on the National Guard Employment Network (NGEN), which is designed to connect guard personnel and their families with employment resources, service providers, and employers. Work for Warriors staff encourage service members to develop profiles on NGEN as a means of building a database of service members seeking employment, and the program uses the database for case management and applicant tracking. NGEN allows the program to connect to Brass Ring, a web-based program that human resource managers use to browse résumés. In early 2014, the California adjutant general released a memo directing commanders to have all underemployed or unemployed National Guard members registered with NGEN by April 1, 2014.[7] Program leadership noted that eventually, the NGEN database may become a means of tracking the employment status of all California guard personnel.

To ensure that guard personnel in search of employment are aware of and make use of Work for Warriors, the program also relies on leadership within units to encourage their service members to sign up for NGEN and provide contact information for the program. When guard personnel are returning from deployments, program staff contact unit leadership to discuss unemployment rates and share information about the program, the website, and the social networking sites. Initially, staff focused on service members a month or two after return, but they have since begun contacting unit leadership before personnel return from deployment. In addition, the adjutant general provides occasional memoranda to communicate his expectation that unemployed service members be encouraged to register with NGEN and access the services of Work for Warriors.

Measurement and Evaluation

Measurement and evaluation are central activities for program staff. All staff are encouraged to regularly collect data and use the data to guide decisionmaking. Program leadership also acknowledged the importance of data in providing evidence on the success of the program and ensuring continued support. The program has never been formally evaluated, though it reports results to the Speaker of the California State Assembly on a quarterly basis, and program leadership testified before the U.S. Congress in March 2013 on the outcomes of the program.[8]

---

[7]   Baldwin, 2013.

[8]   Shepard, 2013.

As noted, job placements are the number one focus of Work for Warriors and therefore play a central role in evaluation. Three teams of resource managers meet weekly with program leadership to share results. The key metrics of focus include jobs posted, résumés submitted, direct hires, indirect hires, and total active jobs the team is trying to fill. The teams also report on jobs of focus, strong new business leads, and any issues that arise. The teams are encouraged to compete on a weekly, monthly, and quarterly basis to achieve the highest numbers of job placements. Program leadership asserts that this competitive spirit drives the resource managers to maximize their results. The program also relies on cost per placement as a key indicator of its effectiveness. Job placements and cost per placement are the metrics most often reported to external parties.[9]

Outside of the team-level breakdown of metrics, program staff reported that NGEN registrants, résumés, applicants, and placements can be broken down to examine success in particular areas and improve the program where issues arise. For example, staff report calculating applicant-to-placement ratios for particular businesses to assess the strength of the partnership and the business's commitment to hiring service members. If the program finds that it is sending many qualified applicants and obtaining few job placements, staff will reach out to the leadership of the business or the business's human resources office to identify the issue and attempt to address any concerns about the applicants being submitted. By holding the business accountable, program staff report that they can ensure that businesses follow through with commitments. Furthermore, the program ends partnerships when it finds that businesses are not strong hiring partners. Another potential use of data to improve program processes is to assess whether certain units have greater or lesser participation in NGEN and to target leadership in units where service members are less likely to be registered (this strategy was mentioned during our visit but has not yet been undertaken).

Measuring social media reach is also an important area of self-evaluation for Work for Warriors. The program tracks likes of its social media pages to assess overall reach. Through Facebook, the program also tracks the popularity of posts according to likes, shares, and reach (number of people who click on the post). In some cases, the staff reported adjusting social media practices according to feedback they receive on posts.

The program has long conducted informal assessments of satisfaction among service members using the program, with a particular focus on individuals placed in jobs and their satisfaction in the placement. The program recently decided to conduct formal satisfaction surveys of program users, and at the time of our visit, the program had recently completed its first satisfaction survey. These survey results are intended for internal use and program improvement. Topics covered include services used, ratings of staff and social media sites, placement success, satisfaction, and likelihood of referral.

---

[9] Shepard, 2013.

Results

Outputs

While the focus of Work for Warriors is primarily on outcomes, the program does regularly track several outputs, including applicants, résumés on file, and business partnerships. Program leadership reported that if they find issues with numbers of job placements, they may begin to require more tracking of outputs to identify the source of the problem. In the first seven months since Work for Warriors received federal funding, the program had 3,217 different applicants. In the three weeks leading up to our visit, the program was averaging 43 applicants a week. The total number of résumés submitted was reported to be 3,508, with an average of 87 résumés submitted per week in the weeks immediately preceding our visit. At the time of our visit, the program reported 1,106 résumés on file and 200 total business partnerships. A weekly metrics reporting sheet that was provided for one of the resource manager teams indicated that the team of three had 19 active jobs they were seeking to fill that week.

Program staff also provided a summary of the social media metrics for the week prior to our visit. Specifically, the program reported approximately 1,800 likes on Facebook and LinkedIn combined, plus 45 followers on Twitter. In the week prior, the program reported achieving 35 new likes on Facebook, 40 new likes on LinkedIn, and 17 new followers on Twitter. In addition to followers, the program tracks reach, or total number of individuals who click on a post, at a total of 4,478. The program also reported 18 Facebook posts throughout the week.

Outcomes

The primary outcomes that are tracked by the Work for Warriors program include job placements, cost per placement, and user satisfaction. The program reported its 1,000th job placement in March 2013, exactly one year after program kickoff, and just nine months after the program reported it was fully staffed. In September 2013, the program had achieved 1,500 total job placements, and by December 2013, the program reached 2,000 total job placements. This suggests that toward the end of 2013, the program was averaging approximately 167 placements per month, or slightly 20 placements per month for each resource manager.

In terms of cost per job placement, the estimates vary depending on the source and timing of the report. Program staff reported a cost per placement of approximately $400 at the time of our visit, while in testimony to Congress in 2013, a figure of $550 per placement was given.[10] These costs are calculated by dividing the total operating costs of the program by the total number of placements. Regardless of the estimate used, program leadership assert that the cost is significantly lower than the $8,000 to $10,000 they estimate it costs on average to fill a job position using a headhunter.

---

[10] Shepard, 2013.

RAND was not able to independently confirm these estimates of headhunter costs or to find comparable cost-per-placement statistics from other sources.

The evidence from the satisfaction survey that the program administers was mixed. Approximately 70 percent of respondents reported that they were "satisfied" or "very satisfied" with the services that Work for Warrior provides. However, 18 percent reported they were "somewhat dissatisfied" or "very dissatisfied," so the program views this as room for improvement. Among those surveyed, 66 percent report that the program had not yet helped them to find a job. However, it is important to note that the target population for the survey was focused only on individuals using their services in the last month, so this does not account for individuals who had already been placed by the program prior to that month, and if individuals only started using the program within a week or two of the survey, it may be unreasonable to expect these individuals to have been placed in a job. The most positive indicator for the program is the 88 percent of respondents who reported that they would recommend the program to a friend.

Impacts

At the program's inception, the primary goal was to reduce unemployment among California National Guard members by 25 percent in one year. To identify the total number of unemployed guard personnel, the program conducted a census and found an unemployment rate of 14 percent in spring 2012. The program reports that these estimates were comparable to Current Population Survey measures from the Bureau of Labor Statistics. Based on these estimates and an estimated California National Guard population of 22,000, the program calculated that the Guard had 3,080 unemployed personnel, and a reduction of 25 percent would require the program to make 770 placements. Given that the program made its 1,000th placement within a year of the program kickoff, program leadership asserts that the program met its goal of reducing unemployment by 25 percent.

Unfortunately, these statistics are not sufficient to determine whether unemployment had in fact been reduced by 25 percent. While the program did likely eliminate 1,000 individuals from the initial pool of 3,080, it is unclear whether new individuals entered that pool during the same time. In addition, it can be challenging to directly attribute declines in unemployment rates to any one program, because there are many changes within the economy over time, and there might be other programs that contribute to the changes in outcomes. The country was experiencing an economic recovery in 2012 and 2013, so unemployment was declining in many states across both civilian and military populations. It is difficult to ascertain what the decline in unemployment might have been in the absence of the program. Program leadership acknowledged the challenges of using unemployment data to measure the impact of the program without a source of comparison.

However, it is important to note that all of this took place prior to the program using federal funding, so these data cannot be used to assess the impact of the program when funded by BYR appropriations. Since receiving federal funding, the program's main goal has been to place as many individuals as possible, and it has been less focused on longer-term measures of impact.

Facilitators and Challenges

Facilitators

According to program staff, there are several facilitators of the Work for Warriors program. Support from the adjutant general and political partners is viewed as an important facilitator, as these relationships ensure continued support of the program and facilitate connections with business leaders across the state. In addition, the high-quality relationships of program staff with businesses, staffing agencies, and other service providers are perceived as having helped the program. In terms of the activities the program carries out, program leadership reports that the singular focus and strong entrepreneurial staff have been facilitators in ensuring maximum success in achieving job placements. Other reported facilitators include the leveraging of free technology, such as social media sites and NGEN, and a commitment from the chain of command to encourage individuals to use the program's services.

Challenges

Work for Warriors has also experienced several challenges. First, program staff report that there are a fair amount of veteran employment programs that are often confused with Work for Warriors, and this confusion can lead to challenges in networking with businesses and with service members. With the California National Guard, the adjutant general's strong support has made it clear that Work for Warriors is the place to go for employment services, which has been helpful in overcoming these challenges. In terms of the relationships with these other employment organizations and veterans' service organizations, the program reported good working relationships with some, yet more-contentious or competitive relationships with others. The more-contentious interactions are largely driven by the notion that these organizations are competing for the same population and the same funding.

Program staff also noted some challenges with those that they serve, including service members and businesses. Among service members, program staff encounter some individuals who are not really interested in finding a job, or who behave unprofessionally when they are given an opportunity to apply for a job. Among businesses, the program faces challenges with getting feedback on the job candidates, and there is a need to constantly assess their business relationships to ensure that the businesses are truly committed to hiring service members. To address both of these challenges, program leadership reports that they focus their efforts on the service members and businesses that demonstrate the greatest commitment to the program.

RAND's Assessment

Whether Stated Goals Are Met

Prior to receiving BYR funding, the program's primary goal was to reduce unemployment among California National Guard members by 25 percent in one year. The program was carefully tracking metrics to determine whether it had met this goal, and program leadership maintains that this goal was met. However, as discussed, it appears that the metrics and analysis design are insufficient to determine whether the program's activities really did lead to a 25-percent decline in unemployment in the California National Guard.

Since the program has received federal funds, its primary goal has been to achieve as many job placements as possible. The evidence does suggest that the program has placed a large number of individuals at a relatively rapid pace, and cost-per-placement figures indicate that the program has been relatively efficient in achieving these results. Based on this evidence, we conclude that the program has largely met its stated goal.

Promising Practices

Based on our analysis of the Work for Warriors program, we have identified several promising practices. First, the program has maintained a clear focus on activities that directly serve the goal of achieving placements, and this singular focus appears to be an important driver in achieving success. In addition, using a high-touch job placement process that supports individuals throughout the entire hiring process (including close involvement with both the service members and the human resources departments in businesses) appears to be a successful practice in achieving high numbers of job placement. Finally, we regard the efforts of the program to track a large number of metrics and use these metrics to drive decisionmaking and program improvement as a promising practice. The program appears to have a clear vision and strong alignment between its mission, its activities, and its outcomes, and that is likely to serve the program well in achieving success.

Areas for Improvement

While we view the program as largely successful, we also noted a few areas for potential improvement. The program has largely focused its efforts in one region of the state and should consider how it might better serve individuals in rural areas and in the southern half of the state. Program efforts to hire an individual who has strong connections in Southern California indicate that they may already be making progress on this front. Another potential area for improvement is to develop a plan for supporting the approximately 60,000 reservists who will soon be added to the program's docket. While the program staff noted that this would be an important area of focus for the next year, it was unclear that there was a solid plan in place to accommodate a quintupling of the size of the population served by the program. Finally, the program could benefit from returning to a greater focus on its longer-term impacts and how these might be tracked more efficiently. The program appears to be having great success with job placements, and it will be important to understand how these efforts are translating into positive long-term outcomes for service members and for California.

CHAPTER THREE

Colorado Marketing and Outreach Program

Bottom Line Up Front

The Colorado Marketing and Outreach Program is among the newest of the BYR-funded programs—at the time of RAND's visit, the program had been operating for about four months. Its primary goals are to assist Colorado National Guard (CONG) Family Program in reaching its target populations and to increase use of its programs. Its activities include revamping the CONG Family Program website, establishing the CONG Family Program social media presence, and providing day-to-day marketing support for the other Family Program offices. Program staff also devote their time to establishing the foundation for long-term Marketing and Outreach Program operations. For example, they developed a handbook and marketing templates, worked with the different family programs to develop an annual marketing outline, and developed needs assessment surveys for different program stakeholders. These efforts help ensure that the program is replicable and sustainable, which constitutes a promising practice that may be useful for other states' programs. The program also is admirable for its use of a wide spectrum of Internet and social media options to market CONG family programs and reach out to new clients.

Because the program was operating in steady state (stable condition) for only one month when RAND's study team completed its data collection (May 2014), we concluded that it had not yet met its goals. Moreover, it is not entirely clear how the program will ultimately demonstrate that it has met its stated goals. Program staff described plans to collect measures of outputs (such as the number of marketing products created and usage of social media outlets), but there was a lack of meaningful measures related to program outcomes and impacts, and no processes in place to collect them systematically.

Overview

History and Goals

Although Colorado has delegated funds for an outreach coordinator in the past, 2013–2014 marked the first year for the more comprehensive Marketing and Outreach Program.[1] As stated in the marketing handbook under development at the time of our site visit, the program's mission is "to create a single resource point which assists the Family Program Offices in reaching their target audiences and increases program utilization by creating established routes of contact to provide consistent, accurate data in a timely manner to the audience in a method of the audience's choosing."[2] By consolidating the marketing and outreach efforts of various Family Program offices (e.g., YRRP, Suicide Prevention, Child and Youth Program) to the greatest extent that is appropriate, the Marketing and Outreach Program office can identify common needs across the programs, conserve resources, ensure consistency in marketing efforts, and provide additional advertising routes. Ultimately, the program is also intended to make the community more aware of guard and reserve families in Colorado and to engage community organizations and vendors to support those families.

Target Population

The Marketing and Outreach Program supports CONG Family Program staff who, in turn, run programs for which guard and reserve personnel from all components, their families, retirees, and veterans are eligible. There seems to be a difference between the eligible population and the target population, however. When asked about the size of their target population, the marketing coordinators estimated the Army National Guard population was roughly 5,000 people, including personnel and their families, and their "supporters" (i.e., extended family and close friends that guard personnel designated as points of contact), which on average is two per person. Air National Guard personnel and dependents do not seem to have been included in this figure. Program staff were unable to provide a size estimate for the reserve side of the overall Reserve Component, noting that reserve personnel and their families are supported through separate resources. That stated, Family Program staff do attend local reserve events, such as YRRP events, so presumably the Marketing and Outreach Program would support marketing for those events as requested. In FY11-related paperwork that Colorado submitted to the Office of the Assistant Secretary of Defense for Reserve Affairs, it indicated supporting 4,500 Army National Guard personnel, 1,450 Air National Guard personnel, and 3,000 family members. This corresponds to a separately reported

---

[1]   Much of the information in this chapter is based on a series of site visits and interviews with Marketing and Outreach Program staff, primarily in February 2014. Additional interviews and sources are noted in subsequent footnotes.

[2]   Family Program Marketing Office, *FY 2014 Standard Operating Procedure and Marketing Handbook*, draft February 2014, p. 3.

estimate of a total target population of 20,000 (assuming the average of two supporters per service member, plus 3,000 family members).

Resources

The Marketing and Outreach Program is administered by two contractors who serve as marketing coordinators. Both have military backgrounds (one is prior military, the other a military spouse) and experience with marketing, and together they offer both Air National Guard and Army National Guard perspectives. Colorado received $190,000 in BYR funds in FY13, and a portion of those funds covered one contractor's salary for one year and the other contractor's salary for three months.[3] The second person's contract was extended via other funding sources. The program's needs for hardware, software, and technical support are met by the State Family Program and G-6 (Office of the Army Chief Information Officer). In general, marketing and outreach products (e.g., brochures, the Family Program website, needs assessment surveys) are supported by other funding sources; BYR funds cover only the contractors' work on them. The Marketing and Outreach Program also benefits from the efforts of the CONG Public Affairs Office, drawing from its news stories and photographs from time to time to support marketing initiatives.

Program Activities

Activities for Target Population

As of our site visit in February 2014, program staff had been working for about five months. Part of their time was spent laying the foundation for long-term program operations by holding kickoff meetings with all Family Program team leads to understand their marketing needs. They also drafted standard operating procedures (SOPs) and a Marketing and Outreach Handbook, which included a marketing request template and marketing After-Action Review (AAR) template. Program staff built a revamped version of the CONG Family Program website, with updated links, new contact information, and additional content. This website replaced an earlier version that was less comprehensive and had been deactivated for an undetermined period of time. Further extending the CONG Family Program's use of Internet-based resources, program staff also established a social media presence on Facebook, LinkedIn, and Twitter and started using Quick Response Codes to integrate their printed and Internet media.

---

[3]   In one of our interviews, we were told that the BYR funds were also used to lease office space. The leased space, close to yet separate from Joint Force Headquarters, permitted staff from all the different family programs to work in close physical proximity, thereby facilitating more collaboration and more-effective case management when needed. We were also told that the separate location offered not only a one-stop shop for clients but also a level of privacy that would likely be lacking in the Joint Force Headquarters building.

Throughout that time frame, Marketing and Outreach Program staff also provided day-to-day marketing support for Family Program office events by designing flyers and magnets, executing email marketing, and applying other marketing tools and products. Program staff also told us about plans to work further with all the program offices to develop annual marketing outlines; to develop and field needs assessment surveys for commanders, Family Readiness Group leaders, and family members; and to use information originally collected for Army National Guard emergency contact purposes to build a target population database for micromarketing.

When we followed up with program staff three months later, we learned that the SOPs and handbook we received in draft form were not signed off on, and instead a memorandum was being developed that would largely replace those materials. This meant that Family Program staff were not yet required to use the Marketing and Outreach Program for their marketing needs. While many of them did, using the Marketing and Outreach Program was optional, and such documents as annual marketing outlines and marketing AAR templates were not being extensively used. With respect to the needs assessment surveys, the commanders' version was fielded at the start of May 2014, so no results were available to share with us. The family surveys and Family Readiness Group surveys were still being finalized, with plans to field them in summer 2014. Finally, the target population database had not been developed, largely due to issues related to changing the database platform from Microsoft Access to Microsoft SharePoint.

Outreach

Also during our follow up, we learned that the Office of the Army Chief Information Officer (G6) was in the process of installing information channel televisions at all the armories, and about half of them had been installed. The televisions will display a rotating series of seven-second information slides. Program staff were creating slides and planned to tailor them for each armory. They were also in the process of revamping and consolidating how emails are sent to the approximately 5,000 email addresses that the Family Program has as a whole. They were using a relatively inexpensive email marketing application called Mad Mimi, which offers ways to create, send, share, and track email newsletters more efficiently. The application is also self-sustaining in that people can add themselves to the list and remove themselves without any human intervention. Marketing and Outreach Program staff were advertising the transition to this new platform, which they called Family Program Connection, at the time of our follow-up and planned to roll it out in summer 2014.

Measurement and Evaluation

The Marketing and Outreach Program has not been the focus of an evaluation, internal or external. At the time of our site visit, program staff were still relying primarily on informal feedback from the Family Program staff, but they did discuss plans to collect more information systematically. As of April 2014, they started creating a monthly report intended to summarize the marketing products they worked on that month. As described earlier in the chapter, program staff spent part of their time working on needs assessment surveys, which could be used to gauge the program's productivity and effectiveness. Also, the Marketing and Outreach Program Handbook includes marketing requests and AAR templates, which, if used consistently, could provide a basis for evaluation.

Other plans for measuring the program's productivity and effectiveness include collecting information that could inform changes to the mix of marketing approaches and products. The planned use of Google Analytics seems especially helpful in this regard because it will enable the program to determine which social media outlets receive the most attention, which specific pages of the Family Program website are accessed the most, and whether they need to do more to accommodate people accessing information using smartphones or tablets. During the feedback session at the end of our site visit, program staff also started brainstorming about ways to improve tracking how recruiters use their products.

Results

Outputs

At the time of our site visit, program staff reported that the number of marketing requests they received in a week was low because they were just getting started and were not yet mandated as the marketing method for the CONG Family Programs to use. Three months later, in May 2014, it was still not mandatory for the other Family Program offices to use the Marketing and Outreach Program, and few indicators of output, in terms of either marketing products developed or client usage of marketing materials, were shared with us. In May 2014, program staff reported that the Facebook group had approximately 200 members, LinkedIn had 22 connections, and the Twitter account had nine subscribers. As expected, the LinkedIn connections tended to be professional contacts, not families. In addition, the program had some information from Google Analytics, but because program staff had only one month of data available, they had not yet done any trend analysis.

Outcomes

During interviews, program staff identified the following outcomes for their program, organized by the type of user:

- service members, dependents, "supporters," and commanders: increased awareness of available programs
- service members and dependents: increased program utilization
- commanders: more-efficient routing of service members and dependents to needed support
- the community: increased awareness of guard families in the community and increased involvement in helping to meet their needs.

Given the newness of the program, there was no information available on results to date. Some of the templates developed and data collection plans that program staff told us about should facilitate collecting data related to some of these measures, particularly those related to awareness. Other measures, such as more-efficient routing to support and increased awareness and involvement within the community, may be harder to operationalize.

Impacts

During interviews, program staff identified the following impacts for their program:

- positive perceptions of Family Programs (by those who would use them)
- cost savings due to reductions in duplicative effort
- increased program efficiency by improved micromarketing ability
- support for commanders' recruiting, readiness, and retention goals.

Again, given the newness of the program, little information was available regarding the program's impact. Program staff did advise us that the Mad Mimi email marketing tool resulted in an annual savings of 850 hours and stated that it would require only about 20 minutes per week to manage it. The savings stem primarily from eliminating duplication of effort. Also, because people can opt in and out of the database, less effort will be required to manage the subscriber list.

Other impacts were hypothetical at the time of our study; program staff still needed to develop ways to measure these impacts and to collect information about them in a systematic way. Further, in the case of program efficiency improvements stemming from micromarketing, the database that would make them possible was not yet in place, so clearly this impact had not yet been achieved. Examples of support for commanders' goals include (1) getting a guardsman to financial counseling before financial problems mount and he loses his security clearance, with negative consequences for unit readiness, and (2) having recruiters use Marketing and Outreach Program materials to show how much CONG does to take care of families. But again, ways to measure and track these indicators of effectiveness had been neither specified nor implemented at the time of our study.

Facilitators and Challenges

Facilitators

Program staff viewed having the CONG Family Program website domain purchased and website in place before they began their work as very helpful. Those setup efforts typically require a great deal of time, and having that taken care of before the BYR funding was received enabled program staff to get the revised website up and running in relatively short order. In addition, having all the Family Program offices in the same office suite as the Marketing and Outreach Program was perceived to facilitate brainstorming across programs and made it easy for someone to come to one of the program staff with an idea before filling out a marketing request form. Avoiding appointments and travel time to different office locations helped to keep the program more nimble and responsive to day-to-day marketing needs.

Challenges

We also observed two main challenges. First, program staff explained that the FY13 funding was delayed until September 2013. As a result, the contracts began on the last day of FY13. The delay in funding meant that the contractors did not have a full year to spend the funds, and as noted, the program was only in operations for about five months at the time of our visit, in February 2014. Perhaps of greater concern, the Marketing and Outreach Program seems to be impeded by a lack of a mandate. As of May 2014, the individual Family Program offices were not required to use the Marketing and Outreach Program. Although some program offices clearly were doing so, this means that duplication of effort still persisted and cost savings were not fully realized. The lack of a mandate also hinders information collection and analysis. For example, if marketing request forms and AARs were regularly submitted, it would be easier to track Marketing and Outreach Program products as outputs in order to gain at least a perception of their utility and to make changes as needed. Some of this was happening informally, but not in a manner that would lend itself to tracking over time.

RAND's Assessment

Whether Stated Goals Are Met

Although the program had been in operation for about five months, as of May 2014, it had been operating in steady state for only one month. Based on this timing issue, we conclude that the program has not yet met its stated goals. Moreover, it remains unclear how the program will measure whether its stated goals have been met now that it is in steady-state operations. The outcomes most directly related to the program's stated goals (to include increased program awareness, program utilization, and impacts, such as the Family Program's client perceptions and cost savings) likely could be measured, but metrics had not yet been developed as of May 2014. Other potential impacts discussed during our interviews, such as support for commanders' recruiting, readiness, and retention goals, will be extremely difficult to gauge, especially given the Marketing and Outreach Program's relatively small scope.

Promising Practices

The program is commendable for its use of a wide spectrum of Internet and social media options, such as Facebook, LinkedIn, Twitter, the new Family Program website, televisions in armories, email marketing software, Quick Response Codes, and Google Analytics. This can help CONG reach its personnel and families using one or more of the media they are most comfortable with (or rely on the most), and the various media can be mutually reinforcing. The email marketing software will help the program do more with less, and Google Analytics can help program staff conduct trend analysis, see which social media are more—and less—effective, and even see what kinds of devices people use to access the Internet.

Also, program staff undertook efforts to make the program replicable and sustainable—two ways to ensure a program is of higher quality.[4] Specifically, the contractors created products (website, social media presence), documentation (marketing request template, AAR template), and processes (annual marketing outline to help maintain marketing and outreach gains even if BYR funds decline or disappear). Some of these are still in a concept stage or are optional rather than mandatory, so their actual utility is still to be determined.

Areas for Improvement

Although all guard and reserve personnel are eligible for the programs for which the Marketing and Outreach Program provides support, the program seems to be not only Guard-focused but even Army National Guard–focused. For example, the program's initial target population estimate was for the Army side of the Guard, and the target population database program staff envisioned for micromarketing will cover only the Army National Guard and its supporters. The program should ensure that the Air Force elements in the state are adequately served, especially when the micromarketing efforts commence.

In addition, if a program is funded for only one year, as was the case for the Marketing and Outreach Program, those tasked with its oversight should ensure a large portion of that time is spent doing rather than planning (that is, a six-month

---

[4] Acosta et al., 2014.

ramp-up time seems excessive). This year was a foundational year in many ways, but moving forward, the program needs to more carefully balance time spent investing in long-term ideas and products versus day-to-day support and quick wins (such as small cost savings).

Finally, as noted, the program needs to do more to develop meaningful metrics related to outcomes and impacts and put measures in place to collect them.

CHAPTER FOUR

Florida Guard Family Career Connection Program

Bottom Line Up Front

The Florida Guard Family Career Connection (FGFCC) program is intended to be a one-stop shop for employment. The program focuses on employment-related activities, including developing résumés, building service member confidence, posting jobs, holding job-related events, and connecting service members to other services. In an effort to increase the pool of jobs available to service members, the program also devotes substantial efforts to networking with businesses.

RAND's assessment is that while the program has set clear goals related to employment, it cannot conclude whether these goals have been met, because the data collected appear to be insufficient. While employment numbers are reported in program materials and by program staff, it is unclear how these numbers were collected, and staff do not appear to have confidence in the data collection efforts or the resulting statistics. Nonetheless, the program demonstrates several promising practices, including building strong relationships with businesses and focusing on businesses that are committed to hiring service members. In addition, the personal relationships with units, service members, and employers help to ensure trust in the program. The main area for improvement is to increase the program's tracking of metrics and to use these metrics for internal program refinement. In addition, the program's activities and strategies are not well documented, and given that the program is largely reliant on a single individual, this lack of documentation raises doubts about the program's sustainability.

Overview

History and Goals

The FGFCC program is intended to be a centralized location to help Florida National Guard members and their families with all employment-related needs. In 2010, the 53rd Infantry Brigade Combat Team returned from deployment with a 38-percent unemployment rate.[1] A January 2011 Florida National Guard survey found that unemployment was at 17 percent. After a history of high unemployment rates among returning service members, the FGFCC program was created to serve as a one-stop shop for guard personnel and their families seeking employment services. The program focuses on connecting personnel to other existing service providers and directly providing services to help personnel and their families secure employment.[2]

The initial goal of the program was to ensure that the Florida National Guard unemployment rate was two percentage points below the state unemployment rate. In recent years, the goal has shifted to a target of 3-percent unemployment or lower among members of the Florida National Guard. A longer-term goal of the program is to build resilience. In testimony to Congress, the adjutant general reported that unemployment is an important factor contributing to an elevated rate of divorce, suicide, and other challenges for members of all military services and components.[3]

Target Population

The primary population that FGFCC has targeted is unemployed guard personnel, and the program has focused most strongly on individuals returning from deployment. The approximate number of Florida National Guard members is between 10,000 and 12,000, of whom approximately half have been deployed.[4] Young service members are the most likely to face employment challenges, so they make up a disproportionate piece of the target population. However, over the past year, as the pace of deployments has slowed, the program expanded to assist unemployed family members and underemployed members of the Florida National Guard. The program will assist other service members (e.g., reservists) if they ask for help, but it focuses on the National Guard.

Resources

The FGFCC program has been supported primarily by BYR congressional appropriations. Program funding started at $110,000 in FY11, and increased slightly to $119,000 in FY14. The program falls under the Manpower and Personnel Directorate (J1). In addition to the federal funding, the program receives donated office space and office supplies from J1; computers and Internet service are donated by the Department of Labor for job fairs and YRRP events at which FGFCC is providing services. While program materials that we received note "state support" for the program, it does not appear that this support was financial.

---

[1]   Much of the information in this chapter is based on interviews with FGFCC program staff conducted in February and April 2014 and materials that the staff provided. Additional sources are noted in subsequent footnotes.

[2]   James D. Tyre, Assistant Adjutant General, Florida Army National Guard, "Putting America's Veterans Back to Work," testimony before the House Committee on Veterans Affairs, June 2011.

[3]   Tyre, 2011.

[4]   Tyre, 2011.

The program is run by a civilian contractor who acts as the program director and sole staff member. The program director was hired at the inception of the program and she is perceived as its driving force. She had prior experience as a corporate head-hunter and had worked with Employer Support of the Guard and Reserve (ESGR) and Hero2Hired (H2H) prior to joining FGFCC. When hiring a director and sole staff member for the program, J1 looked for someone with skills in account development, interpersonal relations, sales, and technology resources. J1 leadership wanted a person with passion for the program, a proven track record in employment issues, and substantial knowledge of the military. The program director is assisted part time by three J1 staff members, who provide assistance with technology and database development, as well as with organizing and staffing job fairs. However, our conversations with staff at the site suggest that the program director could benefit from additional staff assistance. Previously, the program would push job opportunities outward to the state's 16 Battalion Career Counselors, who would then work one-on-one with the service members.

One of the program director's first efforts was to interview J1 staff across all other programs to identify existing employment-related resources and utilize those resources to avoid duplicating efforts. Other J1 programs are therefore important resources to the FGFCC program. In addition, program leadership cited the relationships with businesses as an important resource, because they provide the source of the jobs that service members are placed into. The program director noted that her prior experience as a headhunter and with ESGR and H2H gave her a substantial advantage in the required business relationships.

In addition to relationships with businesses and other programs serving guard members, technology acts as an important resource. The online networking community provided by H2H allows the program director to track service members in the state who have developed profiles and provide them with assistance, as well as allows service members to apply for jobs. The program originally started with an internally developed database and then decided to switch to the H2H portal to track individuals.

Program Activities

Activities for Target Population

The FGFCC program provides a wide variety of employment-related services that are largely delivered one-on-one to individuals in the target population. One of the first efforts is to ensure that individuals are registered in H2H in order to build a database of those needing assistance. Through H2H, the program director can track service members and provide them with the support they need. When individuals register through H2H or otherwise contact the program director, she can help them with job preparation, including résumé development, preparation for interviews, and general counseling and confidence-building. In addition to providing individual employment services, a major activity for the program is to disseminate job opportunities through H2H, Facebook, and other informal channels. Employment events and career fairs are other activities that the program utilizes to help connect service members with employers. The size of these employment events appear to vary. We heard from program staff that the program generally does not feel that large job fairs are effective and therefore focuses on events with four to six employers; nevertheless, in a prior interview, a staff member described an event with 31 employers. Finally, as noted, an important activity for the program is to provide guard personnel with referrals to other services when the program cannot directly meet their needs.

According to the program director, an important aspect of her job is to develop connections with businesses around the state. To strengthen connections with businesses, she networks with employers, attends YRRP events that invite employers, screens businesses to identify those that are interested in hiring service members and have open positions, and encourages employers to post open positions on H2H.

Outreach

The efforts to provide outreach to commanders were noted as an important activity for the FGFCC program. The program director reported that she is constantly working to market the program to both commanders and businesses. To provide outreach to commanders and guard personnel, the program relies primarily on a brochure that is distributed to armories and directs personnel to visit and register for H2H. In addition, a website directs personnel to H2H and provides contact information for the program director, and a Facebook page provides access to the website, links to job postings, and other important information for those seeking employment. Once an individual registers or otherwise contacts the program, the program director reaches out and determines what employment support services might be needed. The program reports that the level of engagement among commands varies. To increase levels of engagement, the program staff would like to create competition between commanders on the number of individuals who register with H2H.

As noted, there are a substantial number of outreach activities focused on businesses. However, the program director noted that in many cases, these efforts are no longer needed because news about the program has spread by word of mouth and employers are now calling her to find out how to partner with the program.

Measurement and Evaluation

The program director reported that she is not conducting any internal measurement or evaluation of the program. In addition, the program has never been formally evaluated. Leadership overseeing the program within J1 reports that they make efforts to track unemployment rates from commanders, H2H, and internal job placement data, but these data are referred to as "fluid." In addition, program staff note that the need to collect more accurate data is an area for improvement. The program is also able to track the number of individuals registered in H2H and the number of employers they have partnerships with. It does seem that the program tracked placements in the past, as several resources report on placement numbers.[5]

Results

Outputs

The only metrics that the FGFCC program tracks with regard to outputs are the number of individuals registered in H2H and the number of employers the program has identified as partners. At the time of our site visit in April 2014, there were 224 service members registered, and the program had a database of more than 150 veteran-friendly employers. Although an October 2012 presentation notes that the FGFCC Facebook page has 90,434 fans, as of February 2015, the Facebook page indicated 772 likes.

Outcomes

According to the program director, the program no longer tracks outcomes. However, a post on the website of Florida Governor Rick Scott reports 436 placements between October 2011 and October 2012, and the 2012 Assessment of Supplemental Funding provided by the program mentions 25 placements between October 2011 and January 2012, with an expectation of 25 per month in future months.[6] In our interviews with the program director, she noted that an important outcome she wants for the program is to ensure that identified individuals gain "meaningful employment"—that is, a career-type job that holds promise for the future, as opposed to a temporary or dead-end job.

Impacts

The unemployment rate for the Florida National Guard is the sole metric examined to determine the impact of the program. A 2012 presentation on the program indicates that between September 2011 and September 2012, the unemployment rate for Florida guard personnel declined from nearly 15 percent to just 6.8 percent, while the state unemployment rate dropped from nearly 11 percent to 8.7 percent. According to program leadership, the Florida National Guard unemployment rate has now dropped to

---

[5] Rick Scott, "Employment Initiatives Bring More Than 400 Jobs to Florida Guardsmen," Florida Governor Rick Scott website, 2012.

[6] Scott, 2012.

just 3 percent. However, the program staff could not provide any data to support that claim. They also reported that there is substantial uncertainty in their estimates, and they are attempting to better measure employment rates with accuracy.

Facilitators and Challenges

Facilitators

During our site visit, FGFCC staff mentioned several facilitators of program success. For instance, the past employment experience and business connections of the program manager were cited as important factors, and the success or failure of the program appears to rely largely on the abilities of this single staff member. The program manager reported that the support she receives from others in J1 also facilitates success. Moreover, program staff indicated that the FGFCC program is the one place for National Guard members to go to for employment, so there is no confusion with other programs.

Challenges

In addition to facilitators, program staff also mentioned some challenges that they face. First, the staff is concerned about perceptions that this is "just another job board" and that there is a lack of understanding among guard personnel that the program provides individualized experiences. In addition, it is a constant struggle to ensure buy-in from commanders; leadership changes often and it can be difficult to make leadership aware of the program and keep them engaged. The program director mentioned that it would be helpful if use of the program was mandatory. And while it is the intention of the program to serve individuals across the state, it has been a challenge to do so with a single staff member. In dealing with employers, the program director described discrimination against guard personnel as a challenge she sometimes deals with, because it can be difficult for some employers to accept the requirements of National Guard service. Finally, program staff mentioned that with a yearly funding cycle, everything becomes focused on the short term, which can be a challenge to program stability.

RAND's Assessment

Whether Stated Goals Are Met

The FGFCC program started with an initial goal of ensuring that the unemployment rate of the Florida National Guard is at least two percentage points below the state unemployment rate, and the program more recently set a target goal of a 3-percent unemployment rate in the Florida National Guard. RAND's assessment is that the data collected by the program are insufficient to determine whether the program has met its stated goal. While data on National Guard unemployment were reported in an October 2012 presentation to indicate that the program had met its initial goal, program staff were unable to provide information on how these unemployment rates were calculated. Currently, the program reports that it uses a combination of commanders, H2H data, and internal placement data to calculate unemployment rates, but program staff were not able to describe the process in detail and did not have any data to corroborate the most recent reports that the unemployment rates have dropped below 3 percent. In addition, program leadership described the measures as "fluid" and did not express confidence in the accuracy of estimates.

Promising Practices

The FGFCC program has several features that RAND considers to be promising practices. First, the efforts of the program director to develop close relationships with businesses and prioritize networking as a means of identifying open positions appear to have played an important role in the program's success. In addition, the focus of the program on employers that are committed to hiring service members helps the program to ensure that its efforts in building business relationships are as effective as possible. According to the program director, another promising practice identified in the early years of the program was to remain engaged with the units, service members, and employers on a personal level, which helps to ensure trust in the program. Finally, the positioning of the program as "the place to go" and "a one-stop shop" for employment needs is good practice because it ensures that service members and employers are not deterred by service members or businesses confusing it with other organizations.

Areas for Improvement

RAND also noted several areas for improvement for the FGFCC program. However, it is important to note that one of the program's main limitations is its small size; with a single staff member for all of the Florida National Guard, it can be difficult to go beyond the provision of basic services. That said, the program could benefit greatly from improved tracking of metrics. Focusing on a single metric that the program admits cannot be tracked with great accuracy limits the ability of the program to use data to evaluate program success and drive program improvement. In addition, it can be challenging to make direct connections between the impact of a program and unemployment rates when the economy is experiencing significant improvement, so it is useful to have other data that indicate how the program may be driving changes in unemployment. Collecting and analyzing data to drive internal evaluation and program improvement are seen as key factors to ensuring success in organizations that deliver services.[7] A related area for improvement is to identify a set of activities and intermediate outcomes for which data can be regularly collected. As a final area for improvement, the program has not documented the activities, strategies, and processes of the program. According to leadership overseeing the program, the success is largely driven by the program director, and if she were to leave, the program would be likely to be difficult to sustain, especially absent such documentation.

---

[7] Howard S. Bloom, Carolyn J. Hill, and James A. Riccio, "Linking Program Implementation and Effectiveness: Lessons from a Pooled Sample of Welfare-to-Work Experiments," *Journal of Policy Analysis and Management*, Vol. 22, No. 4, 2003, pp. 551–575; David Jason Fischer, *The Road to Good Employment Retention: Three Successful Programs from the Jobs Initiative*, Baltimore, Md.: Annie E. Casey Foundation, 2005.

CHAPTER FIVE

Indiana Employment Coordination Program

Bottom Line Up Front

The Indiana Employment Coordination Program (ECP) was designed as a one-stop shop for employment. The program's goals are to provide service members with job-seeking skills and placements in careers, to make sure that they feel supported, and to reduce unemployment and suicide rates in the National Guard. In addition to providing personalized services to individuals, the program posts job announcements and holds employer engagement events to connect service members to potential job placements.

RAND's assessment is that the program is partially meeting its goals. The program's goals are not well defined and lack clear benchmarks. Moreover, in several cases, either the program is not tracking the metrics needed to assess progress toward goals or the data are regarded as unreliable. However, the program is serving as the one-stop shop for employment as intended, and the program is achieving placements. The study team identified several promising practices, including building relationships with employers and other organizations that serve the military and hosting smaller employer events to improve success rates. Areas for improvement include the need for a clear chain of command, the need for specific measurable goals, and the need to use and apply output and outcome data more constructively—specifically for internal program improvement.

Overview

History and Goals

The ECP was created in 2007 with a staff of two contractors from J1's human resources department.[1] At the time, there was substantial concern about suicides among Indiana

---

[1] Much of the information in this chapter is based on interviews with ECP staff conducted in December 2013 and April 2014. Additional sources are noted in subsequent footnotes.

National Guard (INNG) members. Upon further study of the issue, INNG discovered that suicides were largely linked to financial problems, and lack of employment was a substantial driver of financial problems. To address these issues, the adjutant general decided to create a program to help with employment for INNG.

While the program does not have an explicitly defined program goal, program staff suggested that the program has several goals. One staff member reported that the main long-term goal for service members is to enhance career-building skills, such as the ability to search for jobs, market oneself, and network. Several other staff members noted a goal of reduced unemployment rates for INNG members. In the shorter term, one staff member cited the goal as immediate job placement. On the other hand, another staff member suggested that placements were not necessarily a focus and that the program was really created to make sure that service members feel they are supported and have opportunities to improve themselves.

In the same year that the program was created, INNG established a J9, where the program is now housed. Program staff report that the adjutant general decided that there was a need to pull out services that are focused on supporting personnel and families into a single department that is separate from traditional human resources functions. One ECP staff member described J9 as a one-stop shop for the needs of INNG members and their families. At the time of our visit, however, the positioning of the program was in transition. State leadership, including the adjutant general, determined that the program should transition to the Indiana Department of Veterans Administration (IDVA) in 2014.

Target Population

The program has primarily focused on targeting INNG personnel, but ECP will not turn away any service members seeking employment help. Indiana has the fourth-largest National Guard population in the country, with more than 14,000 personnel in the Army and Air National Guard.[2] According to several staff members, there has been some pressure to focus explicitly on individuals returning from deployments, but the majority of the job applicants they assist have not seen combat. Most program staff reported that there are many programs available to help returning service members, so there is not a need to focus exclusively on this population, and ECP should continue targeting all National Guard members—regardless of whether they have seen combat. While not targeting outreach beyond the National Guard, program staff believe that ECP should continue providing employment assistance to all service members.

The program often focuses on the youngest demographic (ages 18–24) because program staff report that these individuals are most likely to be unemployed. ECP efforts are also largely targeted toward larger units and armories because ECP employment coordinators have limited time and must be efficient in reaching as many INNG

---

[2]   Indiana National Guard, "Fact Sheet," web page, undated.

members as possible. In addition, the program is available to INNG family members, though these individuals make up a very small portion of the population that the program currently serves.

Resources

Annual costs for ECP prior to receiving federal BYR funding are unclear. The program was funded with existing J9 staff and resources, with no dedicated state funds set aside to support the program. In fact, funding has been a major issue from the start of the program. ECP was understaffed, and staff had to borrow supplies and could not travel to armories. In FY13, the program received two streams of BYR funding—one for $400,000 and one for $171,000. In addition to the four staff members donated from other J9 programs, the federal BYR funding is supplemented with donations of office space in the J9 building, space at the armories, and office supplies. According to program leadership, the hope is that with the transition of the program to its new home at the IDVA, a more permanent state funding source may be identified.

As noted, the program started with just two contractors, but in 2009, four new technician positions were created. From January through July 2013, the program was reported to be severely crippled, as it had not received the federal funds and could not renew contracts for existing contractors or hire new contractors. In addition, two of the existing contractors were deployed, the program manager went on maternity leave, and the fourth staff member went to Officer Candidate School. When the federal funding was received, the program added seven staff members, for a total of 11 staff at the time of our visit in April 2014. The staff includes a program manager, seven employment coordinators, an administrative staff member, and two additional staff members. The main reported benefit of the extra staff is that they allow each employment coordinator to have more time and resources to focus on a smaller region, developing better relationships with businesses, following up with them, and establishing closer connections to service members in the coordinator's region. In addition to salaries, the federal BYR funding was used to pay for staff travel and training.

The qualifications of the staff vary widely. The education levels range from no college education to graduate degrees, and all of the staff members have varied employment experiences. However, everyone on staff is a current or former service member. Most of the staff agreed that having experience in the military was necessary because staff members need to be able to understand the challenges that service members face. On the other hand, several mentioned that it is useful to present oneself as a civilian to businesses and, in some cases, to units to ensure sufficient respect and avoid being perceived as a recruiter.

The most important quality for staff members appears to be the ability to network. For networking with businesses, important skills include a strong motivation to get out and be creative in making connections and the ability to present oneself appropriately. In accessing service members, the key skills related to networking include the ability to get buy-in from units, persistence in dealing with service members who may not be motivated to get a job, and the ability to connect with the service members and support them.

Initially, there was no training for employment coordinators, and staff report that they were sent into the field to figure it out themselves through creativity. Eventually, the program leadership realized that a little training was needed, such as how to enter data into databases. However, according to one staff member, the training received by employment coordinators has been "very basic."

In addition to program staff, there are several other important resources described in our discussions with ECP staff. Technology has played an important role in facilitating communication and outreach, including using social media to advertise job openings and H2H to track individuals, post jobs, and host the program website. Prior to settling on H2H, the program used several other technology-based options, including NGEN and an internally designed website. ESGR donated several licenses for using H2H.

However, while technology is an important resource, one staff member noted that it is the person-to-person interactions that really distinguish the program, and it is important that technology is viewed only as a tool rather than the central component of the program. In addition, one of the employment coordinators developed a résumé template, and several staff members see this as an important resource in supporting résumé-building with service members.

Relationships are also viewed as important resources. The program has relationships with more than 150 employers, and these relationships are viewed as critical to supporting job placements. Program staff also report relationships with other staff in J9 (e.g., 13 family assistance coordinators) as critical to supporting their efforts by facilitating a two-way referral process between programs.

Program Activities

Activities for Target Population

The job of the program manager is to oversee the contractors, provide outreach to some of the larger companies, and help with the integration of ECP into IDVA. The administrator was previously focused on service member intake and assignment and was responsible for outreach through social media and database management. This staff member has been working part time at IDVA as one means of integrating ECP daily activities into IDVA. Recently, the staff member was assigned a region of the state, so he is acting as an employment coordinator as well.

Employment coordinators are responsible for the majority of activities with service members and employers. According to program management, ECP is designed to service both groups and is often referred to as "two customers, one mission." For activities that are focused on service members, the ECP employment coordinators are intended to be one-stop shops for employment, helping service members to find employment in the short term and build employment-related skills for the long term. To prepare service members for employment, employment coordinators provide help with résumés, prepare job candidates for interviews, and teach job search skills (e.g., converting salaries, identifying career jobs). To place individuals into jobs, employment coordinators post jobs openings on websites, filter job applicants, match employers with service members, attend YRRP events, and host employer engagement events. Employer engagement events are a new activity that is intended to replace traditional job fairs and provide opportunities for three to four employers with multiple open positions to interview and potentially hire individuals on the spot.

According to program management, however, "It always starts with employers." The initial focus of employment coordinators is on networking with businesses to identify job openings and finding employers who really want to hire service members and who have open positions. Face-to-face contact with employers was reported by at least one staff member as being particularly important. The employment coordinators are increasingly circling back with employers, even if there were no National Guard members that applied to their job ads. This helps let the employer know that the employment coordinators are closely monitoring the position and filtering applicants to ensure that the employers receive the best possible INNG candidates. In addition, employment coordinators ask for feedback on the applicants that are considered. Finally, employment coordinators spend time advocating for the program and the skills that service members can contribute to employers.

For the most part, employment coordinators are expected to focus on whatever activities they think are most important and may be best for their region. The employment coordinators we spoke with placed the greatest emphasis on business networking, résumé development, and the screening of candidates. We received conflicting reports of the program's focus on résumés. Some in leadership (particularly those at IDVA) think that there are too many organizations already focused on résumé writing, so the program staff should not duplicate efforts. However, some program staff feel that résumé writing is an essential part of the job when you are trying to get individuals matched with available jobs in a short time frame.

Outreach

Program staff spend a substantial amount of time reaching out to units, businesses, and service members to educate them about the program. Service members primarily learn about the program through the armories. Face-to-face networking with readiness noncommissioned officers (NCOs) at the armories has been important in communicating the importance of the program and how the units can help by providing information on unemployed service members. Other efforts to provide outreach at armories include posting flyers, holding office hours, and attending drill days (though this can be difficult because many employment coordinators drill on the same days). Unit leadership is expected to promote the program to service members and to provide employment coordinators with lists of contact information for unemployed service members. Some program staff reported that it has been a challenge to get the attention of unit leadership and to get good information on unemployed service members, while others indicated that it has not been a problem. While it was previously very difficult to get contact information from the units, INNG leadership has recently begun to provide these lists more regularly. However, several ECP staff members report that it would be helpful for the units to be more engaged.

Employment coordinators place special emphasis on proactive outreach to individuals returning from deployments. The program has started to target units six months before returning from deployment in order to identify which service members will not have employment when they come home. However, as noted, the majority of individuals that the program serves have not been deployed.

Initial contact with service members occurs both from proactive outreach by employment coordinators based on contact lists and through contact initiated by service members who apply for a posted job or call the program's contact number. The other J9 programs often provide referrals to ECP as well. When individuals initiate contact, they are assigned to the employment coordinator in their region, who will then follow up with the service members. Communication between employment coordinators and service members usually occurs through telephone calls, texts, and emails, though there are sometimes in-person meetings. Some staff members report that they will attempt contact with each individual who is referred to them, and if individuals are not responsive after several tries to contact them, they will be dropped from the list. To provide outreach about current job openings, the program uses Facebook and H2H.

Measurement and Evaluation

Indiana's ECP collects a substantial amount of data and tracks a variety of metrics to assess its progress. In order to track activities and immediate outcomes, employment coordinators use a common tracking sheet that feeds into a master spreadsheet held by the program manager. The tracking sheet includes

- résumés edited
- résumés sent
- employer outreach (number of visits to employers)
- walk-ins
- postings on the ECP websites
- service member outreach events
- placements.

In addition to placements, the program has recently begun tracking data on salaries associated with placements. Staff members report that placements can be particularly difficult to track, because oftentimes counselors do not hear back from a person and do not know if they were hired for the position for which they applied.

In addition to tracking data through spreadsheets, the program tracks individuals through H2H. In H2H, coordinators can determine which individuals are actively applying to jobs and where they are applying. However, according to program staff, it remains a challenge to make sure that service members in the system do not fall through the cracks. In addition to service-based activities, the program tracks data on its social media metrics (such as likes, the number of individuals a post reaches, and sharing of the information).

The program also collects data on impacts. To measure unemployment rates, the program collects monthly roll-ups of unemployment data from units. However, program leadership reports that the program would like to improve its efforts to track unemployment.

According to program leadership, the primary purpose of collecting metrics is to push the data upward in order to solicit additional funding. The program has not been formally evaluated, and it does not commonly use data for internal program assessment and improvement. Once data are collected, the program sends monthly updates to J9 leadership and provides additional reports upon request. A sample report indicates that the primary metrics reported are outcomes—including placements and salaries associated with those placements. The reports to J9 leadership also note special accomplishments and challenges that the program may have experienced within a given month.

Program staff generally feel that the metrics they track do not capture a big portion of what they are doing, because much of what they do is perceived as intangible (e.g., networking, meeting with service members' families, helping to comfort a service member). Among the staff that we spoke with, there was general skepticism of the ability of metrics to capture the work that employment coordinators are really doing. According to one program staff member, one of the goals is to "give a warm and fuzzy [feeling] to people and let them know someone cares about them," and this really can't be measured. The program does not encourage job placement quotas for employment coordinators or competition among them as an effort to avoid "micromanagement." There is also a feeling that employment coordinators have different regions that may drive different results.

Program staff reported that IDVA places a much greater (and singular) focus on job placements, and this is an area of concern for some program staff who feel that this focus takes time from activities they believe are important, such as helping prepare résumés and talking with service members to understand the issues they may be facing.

Results

Outputs

The program's range of activities with service members and businesses results in outputs that include résumé editing and submission, walk-ins by service members, visits to armories and outreach to businesses, events for service members, and social media activity. An internal metric tracking sheet identifies outcomes for the 2013 calendar year, though the final quarter of 2013 best reflects the impact of the program because this is when additional staff were brought on and the program recovered from substantial challenges with staffing and funding. In 2013, ECP processed 1,066 résumés, with 388 processed in the last quarter of the year. Assuming that the seven employment coordinators are primarily responsible for résumé processing, this averages approximately 14 résumés processed by each employment coordinator in a month. Of these processed résumés, the program submitted 595 résumés to employers in 2013— including 313 in the last quarter of the year. According to data from the last quarter of 2013, employment coordinators were averaging approximately 11 résumé submissions each month.

In addition to résumés, the program tracks other services provided to service members, including walk-in consultations, outreach events, and job postings on the program website. Walk-in consultations were not tracked until September 2013, and in the last quarter of 2013, employment coordinators reported 29 walk-ins, or approximately one per employment coordinator per month. The program staff reported that 53 outreach events to service members took place in 2013, with 27 of these events occurring in the last quarter of the year. In addition, employment coordinators posted 229 times on the program website, with 86 of the posts occurring in the final quarter of the year. Based on data from the final quarter of 2013, the program was averaging approximately three posts per employment coordinator per month.

In addition to metrics related to services provided directly to service members, the program provided results for 2013 on employer outreach (visits or calls made to businesses) to identify openings and provide outreach about the program. The 2013 metric tracking sheet indicates that program staff engaged with employers 500 times, with 218 of the engagements occurring in the last quarter of the year. These data suggest that employment coordinators are providing outreach to employers approximately eight times each month.

Lastly, the program staff provided us with metrics on social media from the week immediately preceding our visit in early March 2014. The metrics indicate that the ECP Facebook page had 199 total likes, with 23 new likes in the week for which metrics were provided. The posts reached 221 individuals in that week, an increase of 46 percent from the previous week according to the metric tracking sheet. Posts to the ECP Facebook page in that week received nine likes, two comments, and 17 shares.

Outcomes

Program staff report that there are several outcomes that the program would like to have. These intended outcomes include job placements, improved job search skills (including networking, résumé development, and interviewing), a reduction in financial issues, and a perception among service members that someone is looking out for them. However, among these outcomes, the only one that is regularly tracked by the program is placements. In 2013, the program reported 159 placements, of which 81 occurred in the last quarter of the year. Data from the final quarter of 2013 suggest that the average employment coordinator is placing approximately three individuals each month. The program also tracks the salaries associated with those placements. The 159 placements account for nearly $3 million in total salaries, for an average of approximately $19,000 per placement.

Impacts

The program has goals related to long-term impacts both for individual service members who the program serves and for the INNG population as a whole. At the service member level, the program wants to ensure that individuals end up in promising careers, as opposed to jobs that provide little promise for the future. For the INNG population, the program intends to reduce unemployment and suicide rates. Of these three impacts, the program collects regular data only on unemployment, though it may be the case that salary data can provide some measure of career potential among job placements.

  As previously noted, unemployment data are collected through monthly reporting from the units, and there is some uncertainty about the quality of data. For example, at one time, an unemployment rate of 23 percent was reported for INNG, but when program staff started to dig into this, they found that many of these personnel were going to school, some were stay-at-home moms, and others simply were not interested in getting a job. After excluding individuals who were not actively seeking employment, the program estimated a real unemployment rate closer to 6 or 7 percent for INNG. While program staff reported that unemployment data are tracked regularly, we were not provided with any specific data on unemployment that can be used to report on the program's impact.

Facilitators and Challenges

Facilitators

With regard to facilitators, program staff report that being placed in and supported by J9 is extremely important because J9 has an exclusive focus on service member and family well-being. The partnerships with the other programs in J9 are also helpful. In addition, program leadership reports that INNG's emphasis on the program—in particular, the personal support of the adjutant general—has been a major facilitator in ensuring the continued existence of the program, despite funding challenges. Program staff also report that their more than 150 partnerships with employers and other service programs, such as Job Ready Vets, help to facilitate the success of the program. Finally, ECP staff report that technology has been helpful when it is easy for staff to use.

Challenges

There were also some challenges noted by ECP staff. These challenges generally fall into four categories:

1. issues with funding and organization
2. a lack of commitment to the program from employers and service members
3. logistical issues in delivering services
4. challenges with recognition and buy-in from key stakeholders.

With regard to funding and organization, the lack of a clear funding source has been a challenge from the inception of the program. According to several staff members who were with the program before it received federal BYR funding, there was a feeling that they needed to constantly fight to be recognized and they needed to piece together small bits of funding to support what they believed was an important program. Even after receiving a commitment of federal BYR funding in FY13, the program did not receive the funding until late in the fiscal year, leading to challenges in sustaining the program. In addition, the relatively short one-year terms for federal BYR funding are challenging because continuity in the team is believed to be important, but contractors are concerned about job stability. Several staff members reported that fears of job loss among contractors lead to lower levels of commitment to the program.

In addition to funding challenges, the transition of the program to IDVA has been extremely challenging for some staff members. There is a perception of substantial infighting, and some staff members report that the chain of command is unclear. Some staff members suggest that a strong chain of command for the program is needed because it allows for a clear plan to be laid out for the program and ensures that staff are proactive rather than reactive. Differences in viewpoints on the key areas of focus have led to disagreements, discomfort, and a lack of coordination between IDVA priorities and the activities that program staff are engaging in.

Regarding service member commitment, program staff report challenges with service members who have a sense of entitlement and believe they are owed high-paying supervisory positions despite qualifications. The program must work with these individuals to manage expectations and guide them toward more-effective job-seeking attitudes and behaviors through skill-building in such areas as job etiquette. In addition, the program faces challenges with service members who are not willing to actively pursue employment.

The program has also faced challenges with employer buy-in and commitment to the program. One staff member reported that initial meetings with employers can be challenging because employers are skeptical and think that they may be required to pay something, or that the program staff member is a military recruiter. To overcome challenges with employer buy-in, some staff members report that it is useful to present oneself as a civilian, and it is important to develop networking skills to be able to talk with employers as equals.

Among employers who are interested in partnering with the program, additional challenges can arise. Program staff reported that there are many businesses that claim they want to hire veterans but do not actually have a commitment to placing individuals in jobs. The program has started to be more selective about who they invite to employment events to ensure that their efforts will indeed result in job placements.

Some staff members also reported logistical challenges. For example, scheduling with service members at armories can be a challenge because service members may not be able to attend employment events on the weekends because they are drilling. In addition, program staff would like to meet with INNG members on drill weekend because they can connect with a large number of individuals at once. However, many of the program staff members are also in INNG and are unable to visit on drill weekends because they are also drilling. On top of scheduling challenges, one staff member also noted that it can be particularly difficult to provide outreach and services to individuals in rural areas.

Technology has also been a challenge for the program, which has cycled through three different databases in just two years. The prior databases presented challenges that include limited ability to track individuals through the system (their internally created site) and difficulty in ease of use (H2H). However, the program staff reported that this challenge has largely been overcome.

Finally, according to program staff, ECP faces challenges with recognition and buy-in. There is a concern that there are too many other programs focused on veteran employment, and it was initially a challenge to determine where the program fit it. The program has worked to partner where services already exist and fill in where no other programs are providing services. For example, the program has largely focused on INNG members who are not combat veterans because fewer services are available to these individuals. Another challenge is that not all organizations that provide assistance to service members and veterans have been open to developing partnerships with ECP. Some organizations feel like they do not want to share contacts and feel competitive over job placements.

Some staff members also indicated that recognition and buy-in from unit leadership has been a challenge. Several staff members noted that some senior officers confuse the program with other programs, and they do not always understand what the program is or what value it provides to their personnel. In some cases, there may be a hesitancy to shine a light on a problematic issue like unemployment. As a result, some staff members cannot always get the employment and contact information they need, and unit leadership does not encourage individuals to utilize the program's assistance. However, there were several seemingly more connected program staff who did not report recognition and buy-in from the units as a problem, so these challenges may be overcome by individuals with strong networking skills and engaging personalities.

RAND's Assessment

Whether Stated Goals Are Met

The Indiana ECP has a variety of goals that touch on activities, outcomes, and impacts, including

- serving as a one-stop shop for employment
- achieving placements
- reducing unemployment
- addressing financial issues
- reducing suicide rates
- making sure that service members feel supported.

Of these goals, the program is tracking metrics that shed light on only a few, and the program has not identified any benchmarks against which to measure progress. RAND's assessment is that the program is partially meeting its stated goals; the program does appear to serve as a one-stop shop for employment, and the program is achieving job placements—particularly since receiving federal BYR funding. However, without clear benchmarks for job placements, it is unclear whether the program is efficiently and effectively placing individuals. Metrics for unemployment are also being tracked on a monthly basis, but program staff are uncertain about the quality of data.

Promising Practices

In our assessment of ECP, several promising practices were identified. For instance, program staff reported that strong relationships with employers are key to the program's success. Employer relationship-building helps the program identify job placements, knowledge of employer needs can help to better prepare individuals for interviews, and advocacy for job applicants can help to ensure that employers devote a small amount of additional attention to the applications that come from ECP. In connecting employers with service members, program leadership has adopted a new practice of holding employer engagement events that focus on three to four employers who have open positions at the time of the event and a commitment to filling those positions with service members. The program will no longer hold traditional job fairs, because these are viewed by program leadership as ineffective. In addition to ensuring that employers are ready to hire, the advantage of smaller events is the ability of employment coordinators to better prepare service members for the specific employers that are there to hire.

The ability of the program to connect service members and employers with a real person is viewed by program staff as an important promising practice. According to program staff, connecting with a person ensures that service members receive personalized help, and employers are more likely to develop a partnership with the program when they feel a personal connection. Several staff members mentioned that face-to-face contact is particularly useful.

Program leadership also recommended that programs engage in partnerships with other organizations early on. In hindsight, program leadership would have liked to have established a protocol for building and maintaining relationships with other organizations.

Program staff also identified promising practices that were learned from other states. The promising practice that ECP reports learning from Washington state was the need to have more employment coordinators to fully cover the state. Adding staff to the program allowed ECP to have smaller regions for employment coordinators to focus on, and the employment coordinators report that clearly defining these territories is helpful. From Texas, ECP staff learned that it is important to have a database to track service members. As a result, the program shifted from their internally designed web portal to databases provided by NGEN and H2H. The promising practices drawn from Maryland's program include using the Internet to provide outreach and using existing programs to the degree possible rather than trying to reinvent the wheel.

Areas for Improvement

We also identified several areas for potential improvement for ECP. For instance, the program is in need of a stable funding source and a clear chain of command. During the transition from J9 to IDVA, many of the program staff members have become dissatisfied and are unsure who is in command and what they are supposed to be doing. According to reports from program leadership, the vision of the adjutant general is that the transition of the program to IDVA will ensure a permanent home for the program and potentially a state funding source. Program leadership reports that the program will be vastly improved if there is a sustainable funding source that allows for more continuity in staff and programming. The history of the program has been characterized by a lack of funding and concerns about sustainability. Regardless of whether the program increases continuity, it is important that employment coordinators receive better training. The approaches of employment coordinators varied widely, and while variability was viewed as a positive feature, it seems that employment coordinators were also seeing variation in results, and several staff members expressed a desire for more training. With a program that is experiencing substantial transition and a lack of stability, it may be particularly important to create SOPs or, at minimum, provide common training to ensure that the program maintains continuity.

The program may also benefit from a more focused set of goals and a set of activities and outcomes that clearly connect to those goals. The descriptions of the program's goals and activities varied substantially across program staff, and there appear to be no officially stated mission or goals for the program. The goals that were stated lack clear benchmarks, and there appear to be insufficient data to measure progress on several of the goals. It is important that the program's activities, outcomes, and impacts can be measured to ensure that the program is effectively and efficiently meeting the needs of the target population. Rather than viewing metrics as useful only for program justification, the program could benefit from regular internal use of metrics as a means of identifying areas of success and areas where improvement is needed so that refinements can be made to capitalize on the program's successes.

CHAPTER SIX

New Hampshire Care Coordination Program

Bottom Line Up Front

The Care Coordination Program (CCP) aims to identify unmet needs of service members, their family members, and veterans and provide support to meet those needs. The ultimate goal of CCP is to cultivate individuals who are self-sustaining, able to meet their own needs. This typically entails referring program participants, or clients, to relevant agencies that can provide long-term support, but in some cases, program staff provide direct support (e.g., transportation to job training, immediate counseling support). The program provides support in multiple areas but emphasizes suicide prevention, mental health, employment, and homelessness.

CCP appears to have met the goals it measures related to suicide prevention, access to mental health care, employment, and homelessness. In addition to measuring the number of clients it has helped in each of these areas, CCP has also documented that it successfully increased its cost efficiency over time. However, while it has an extensive reporting system to track data, there are additional outcomes that it is not currently measuring. Promising practices include building relationships with clients based on a unique model, hiring program staff who have preexisting relationships with social service organizations, and maintaining an extensive network of resource providers to which CCP clients are referred.

Overview

History and Goals

Easter Seals, a community-based nonprofit organization, began supporting the New Hampshire National Guard (NHNG) through its Veterans Count initiative, which provides philanthropic support for veterans, service members, and their families.[1] In 2007, Easter Seals approached NHNG to help close the gaps in services to veterans by

---

[1] Much of the information in this chapter is based on interviews with ECP staff conducted in January 2014 and February 2014. Additional sources are noted in subsequent footnotes.

leveraging its existing relationships with other nonprofits throughout New Hampshire. Prior to the Veterans Count program, stakeholders felt that there were gaps in services provided to veterans years after returning from deployment and that many service members and veterans were overwhelmed by the resources available to them and unsure of where to seek assistance. Then, when DoD requested that New Hampshire pilot the Joint Services Assistance Program to provide support and services to guard personnel, J1, the New Hampshire adjutant general, New Hampshire Department of Health and Human Services (NHDHHS), and Easter Seals suggested that DoD allow New Hampshire to pilot a program more fitting to a state without any active-component military installations. DoD agreed and began funding the New Hampshire pilot CCP for one year on a small scale in 2007. The pilot program served 12 cases over its first nine months and concluded that 90 percent of the emergency support the program provided could have been predicted and proactively delivered to service members pre-deployment, and 80 percent of service gaps could be filled by existing resources. Given those findings, the program continued and served as the genesis for the current CCP (although the funding sources and where it was housed have since changed).

Since CCP's inception, it has been run in partnership with NHDHHS. However, when the program goes to a federal contract in 2015, NHDHHS will be dropped as a partner. While the program initially fell under the Joint Family Support Assistance Program, it has since become funded by BYR. The program was originally called the Deployment Cycle Support Care Coordination Program, but "Deployment Cycle Support" was dropped in 2014 because the program provides support that is larger in scope than just during the deployment cycle.

The overall goal of CCP is to meet the needs of service members, veterans, and their families, ultimately cultivating individuals who can be self-sustaining. With no active-component military installations in New Hampshire to support service members, veterans, and their families, CCP sees itself as building a "Fort New Hampshire" to serve its target population. CCP emphasizes meeting any immediate needs of clients and then focusing on longer-term needs after building relationships with them. While CCP meets a variety of client needs, the program focuses on goals related to suicide prevention, access to mental health care, employment, and homelessness. CCP does not provide specific numerical goals for these four focal areas. CCP staff identified the importance of cost outcomes, and the program tracks the costs and cost savings it creates by working with clients.

Target Population

CCP serves New Hampshire service members, veterans, and their families. Program staff estimated that there are 5,100 guard personnel in New Hampshire, but they were unsure of the number of veterans residing in the state. The program supports service members and their families before and during deployment. Additionally, CCP has continued to hone its practices to ensure that it is effectively providing support to service members who are not in the deployment cycle. While the program focuses on post-9/11 members of NHNG, it also serves pre-9/11 veterans, as well as veterans and service members from other military branches. Furthermore, CCP recognized that while it has effectively engaged the New Hampshire Army National Guard, as well as more-recent veterans and current service members, one of CCP's goals for FY14 was to engage other branches, including the Air National Guard, and veterans of all eras more effectively. (See "Outputs" in the Results section for a breakdown of the branches served in FY13.)

Resources

CCP is overseen by the director of NHNG's Service Member and Family Services (SMFS) program and the bureau chief for NHDHHS's Community-Based Military Programs. However, in the next year, CCP planned to drop NHDHHS as a partner and move to a federal contract. The SMFS director felt that the current contract made it difficult to manage the program, with NHDHHS serving a middleman function, and that moving to a federal contract would reduce "administrative burden." Easter Seals, on the other hand, expressed that the current contract at the state level allows them some flexibility in expanding and contracting staff based on need.

Paid Easter Seals staff carry out the contract, which includes ten care coordinators and five other staff roles (i.e., director, clinical director, intake director, office manager, and data process and outcomes manager), which some of the care coordinators also fill. Care coordinators have master's degrees and, with the exception of one, are trained in a range of clinical specialties (e.g., marriage and family, mental health, social work). While not all CCP care coordinators are licensed, they all have worked as community caseworkers and thus have experience in building relationships with clients. When NHNG deployed large numbers of personnel, Easter Seals contracted out, worked with, and trained members of the local social service community to meet the increased demand. While some CCP staff work out of the Easter Seals headquarters building in Manchester, New Hampshire, the care coordinators work from home and conduct business using smartphones and laptops.

Numerous Department of Veterans Affairs (VA), military, and community programs and organizations also support the work of the care coordinators by providing the support services to which CCP participants are referred. The care coordinators also rely heavily on personal connections with and knowledge of various resources to help serve participants. CCP and the care coordinators are well tied into social service programs in New Hampshire, which facilitates referrals and deep local knowledge that clients may not be aware of. CCP also has private and state connections that provide for flexible additional funding, giving CCP an advantage in serving the community.

CCP received $889,000 of BYR funds for FY13. BYR appropriations go from the federal government to the Joint Services Assistance Program, which contracts with NHDHHS, which then contracts out to Easter Seals. Additional financial resources available to all service members are at Easter Seals' disposal, including the Veterans

Count funds and Chaplain Emergency Relief Fund; this structure provides direct financial support to service members and allows for flexibility in use of funds.

Program Activities

Activities for Target Population

Care coordinators conduct case management and work with service members, their families, and veterans to meet their needs. Care coordinators begin with a needs assessment that identifies areas in which the participants need immediate help, and then they develop a more comprehensive assessment plan that focuses on developing self-sufficiency. The care coordinators then refer the participants to relevant agencies that can provide long-term support. In some cases, the care coordinators provide direct support to CCP participants (e.g., provide transportation to job training, provide immediate counseling support). Care coordinators also work with a service member's family throughout a deployment and link them with any needed resources.

In addition to serving CCP participants, care coordinators work to build relationships with support services within the state and participate in statewide initiatives that focus on issues that affect the military and veterans. CCP has a representative on multiple state committees (e.g., a drug and alcohol committee, a posttraumatic stress disorder committee), and program staff work with a variety of organizations and initiatives that support veterans, such as Harbor Homes (a nonprofit organization that focuses on relieving homelessness in New Hampshire).

Outreach

CCP conducts outreach to service members, their families, and veterans by giving presentations at predeployment drills and YRRP events. Care coordinators repeatedly reach out to service members or families until they successfully form a relationship. Additionally, they reach potential participants through referrals from other sources and programs—including chaplains, the VA, and job and resource fairs at colleges and universities. In our interviews, program staff noted that word of mouth has been one of their best methods of outreach because CCP has high-profile private partners that provide donations through the Veterans Count initiative. Program staff did not mention any use of social media to conduct outreach.

Measurement and Evaluation

Care coordinators complete several forms as they work with CCP participants (e.g., intake form, referral form, clinical checklist). All of the information from those forms goes into a linked database that staff use to create reports. CCP creates monthly, quarterly, and annual reports that track its outputs and outcomes of interest. In these reports, CCP measures the following outputs: the number of cases served, the hours spent by the care coordinators working with cases, and the number of referrals made. It breaks down these indicators by participant demographics (e.g., gender, marital status, military branch), acuity of case (minimal, moderate, or intensive needs), type of referral (e.g., mental health, employment), and community linkages (where referrals were made). CCP also tracks cost outcomes (e.g., cost per case, cost efficiencies), as well as the amount of philanthropic support that the Veterans Counts program received (because that program is a large source of support for CCP). In addition, CCP gathers more-qualitative information about cases that illustrate specific outcomes but does not enter that information into its database.

CCP looks specifically at care coordinator productivity, cost per case, and specific outcomes (e.g., mental health, employment, homelessness) to consider whether or not it is meeting its goals. However, the program does not provide specific numerical goals for any of these areas. It uses the data to determine areas of support that care coordinators need to target and to provide resources to care coordinators who may need extra support. CCP also does trend analyses with indicators of usage (e.g., types of resources used, use by geographic area) to inform what resources it might focus on (e.g., homelessness) and if it needs to decide where to locate a new care coordinator.

Results

Outputs

CCP has served 2,314 cases (service members, their family members, and veterans) since the beginning of the program in 2007. Of the 523 cases that CCP served in FY13, 201 were service members and their families in the deployment cycle (i.e., predeployment, deployed, or one year postdeployment) and 322 were not (i.e., veterans, service members who are not in the deployment cycle, and their families). In FY13, 70 percent of participants served were in the Army National Guard, 15 percent were full-time military in the Active Component, 7 percent were Army Reservists, 2 percent were Marine Reservists, 2 percent were Naval Reservists, 2 percent were Air National Guard personnel, and 2 percent were unknown. CCP does not disaggregate the number of cases in and out of the deployment cycle by branch of the military.

In FY13, care coordinators made 1,385 referrals, broken down as follows:

- NHNG Family Program (93)
- NHNG Chaplain Emergency Relief Fund (65)
- ESGR (89)
- VA Medical Clinic (94)
- Veteran Center services (66)

- Military One Source (42)
- NHNG Transition Assistance Advisor (45)
- financial counseling (94)
- community mental health programs (73)
- children's services (92)
- employment services (74)
- ancillary services (e.g., faith-based programs, food pantries) (558).

CCP used more than $250,996 in local philanthropic funds to support CCP participants in FY13. The program found that the cost of each case continuously decreased each year from $3,508 in FY08 to $1,496 in FY13.

Outcomes

CCP measured all of its stated short-term outcomes in terms of the number of cases it served in its areas of emphasis—suicide prevention, access to mental health care, employment, and homelessness. In FY13, CCP reported intervening in ten cases where participants were at risk of suicide. As of CCP's FY13 annual report, no CCP participants had died by suicide since the program's inception. In FY13, 10 percent of CCP participants who were service members, 7 percent of adult family members, and 4 percent of child family members received treatment for a previously unidentified mental health issue. That same fiscal year, care coordinators helped 77 program participants secure employment and 40 out of 43 homeless participants secure permanent housing. Housing stability plans were created for 151 participants who had significant risk to housing stability, and 96 of those cases received Veterans Count funds to help ensure stable housing.

While CCP reported qualitative, anecdotal evidence of its longer-term outcomes of building relationships with participants and developing self-sustaining service members, the program is not collecting or analyzing quantitative data related to those outcomes.

Impacts

CCP's intended impacts are improved military readiness and increased understanding among the community of the issues facing service members. At the time of our site visit, CCP was not measuring either of those impacts. Program staff noted that in the long term (20–30 years), they expect there to be cost savings to VA and other social services because CCP is helping meet the needs of service members, their families, and veterans as the needs arise. Staff contrasted this with the lack of services for Vietnam veterans when their needs arose, which required VA and other social services to address those needs much later, at significant costs. While CCP is not measuring long-term cost savings, it is measuring the number of clients that are served in a variety of areas that might eventually require VA resources if not addressed (see Outputs subsection).

Facilitators and Challenges

Facilitators
Many of the factors that facilitate CCP's success have to do with the people and organizations that work with the program. Multiple staff members noted that having a program manager (i.e., the SMFS director) who understands what the program is trying to accomplish is a strength. Many staff members also attributed CCP's success to its support from the NHNG chain of command. Furthermore, the program has good working relationships with other organizations and military programs that provide support—including the Chaplain Emergency Relief Fund and the VA. The fact that Easter Seals operates the program also facilitates success. Easter Seals' resources (e.g., Veterans Counts funds) provide care coordinators with more flexibility in serving clients, and the cooperative agreement between Easter Seals and CCP allows for flexibility in providing services to a range of groups, not just service members in NHNG.

Challenges
Many of the challenges that CCP faces stem from the current structure of the contract and the changes that will occur in the next year. CCP staff noted that they receive little guidance from DoD on how BYR funds can or should be spent. They also identified the uncertainty of funding as a hindrance to long-term planning. Where the funds come from dictates who CCP can and cannot serve (e.g., DoD funds should not support the Coast Guard). In the next year, as CCP moves to a federal contract and there is a bid for services, a new vendor (i.e., not Easter Seals) could win the contract, and the institutional knowledge of the current care coordinators could be lost. Moving to a federal contract during 2015 also means less flexibility and a lack of clarity on the future population that CCP can serve.

RAND's Assessment

Whether Stated Goals Are Met
While CCP met all of its measured goals, some outcomes suggested as overall program goals are not included or measured (i.e., long-term outcomes, impacts). While CCP does not have specific numerical goals (as detailed in the Results section), it is clear that the program is supporting numerous New Hampshire service members, their families, and veterans, especially in its focus areas of suicide prevention, access to mental health care, employment, and homelessness. Furthermore, CCP documented that it has successfully increased its cost efficiency over time.

Promising Practices

CCP credits its success to its model of building relationships with clients and breaking down barriers to services. Furthermore, Easter Seals has relationships with people and organizations in each of CCP's focus areas to help the program achieve its objectives and meet the needs of participants. The care coordinators have created working relationships and have an extensive network of services offered in the state. In many cases, the care coordinators were previously community caseworkers with preexisting relationships with social service organizations. They further build networks and relationships by participating in a variety of statewide initiatives related to CCP's services (e.g., mental health).

Areas for Improvement

CCP staff indicated that the program has some difficulties capturing data to measure some of its stated outcomes—especially long-term outcomes and impacts. In particular, while program staff were interested in client satisfaction, they struggled to collect the data needed to track that measure. The program created an exit survey for clients whose cases were closed, but they received few of the surveys back, and program staff indicated that they were not sure the survey effectively captured whether or not a client's problems were resolved.

CHAPTER SEVEN

North Carolina's Integrated Behavioral Health System, Education and Employment Center, and Legal Assistance Program

Bottom Line Up Front

North Carolina utilized FY13 BYR funds for three programs: the Integrated Behavioral Health System (IBHS), the Education and Employment Center (EEC), and the Legal Assistance program. The goals of IBHS are to assess service members and their families for immediate behavioral health needs, offer therapeutic support, provide case management services, and provide referrals to federal, state, and local resources. IBHS is meeting these stated goals. IBHS has developed a robust data collection system, and it routinely uses the data that it collects to improve the program and refine its SOPs. IBHS has become the primary behavioral health resource for service members in North Carolina.

The goals of EEC are to establish ongoing relationships with veteran-friendly employers in North Carolina and place clients in positions that help them achieve their career goals. RAND's assessment is that EEC is partially meeting its stated goals; however, this is primarily because the program is very new. EEC is meeting its short-term goals of establishing ongoing relationships with veteran-friendly employers and helping service members find employment. However, it is unclear to what extent the program is meeting its other formal goal of placing clients in positions that help them achieve their career goals.

The goal of the Legal Assistance program is to provide service members with access to legal services for a reduced cost or no cost. RAND's assessment is that the Legal Assistance program is meeting its stated goal. The program has helped thousands of service members in North Carolina resolve their legal issues. In addition, the program has saved service members hundreds of thousands of dollars in legal fees and taxes.

North Carolina's Integrated Behavioral Health System

Overview
History and Goals
The goals of IBHS are to assess service members and their families for immediate behavioral health needs, offer therapeutic support, provide case management ser-

vices, and provide referrals to federal, state, and local resources.[1] After a heavy brigade combat team from the North Carolina Army National Guard returned from deployment with 4,000 personnel in 2010, the North Carolina director of psychological health (DPH) was inundated with referrals for mental and physical health issues and at least three hospitalizations a week for nine weeks. This showed a need for a program that addressed the behavioral health needs of service members in North Carolina. The DPH approached the adjutant general, and together they developed a plan to meet those needs. IBHS was formally stood up in November 2010.

Funding for IBHS was initially difficult to procure. First, the program applied for grant money but was unsuccessful, and then the state government funded the program for six months to test it. The BYR funds that the program has received since 2011 have allowed it to develop, identify what is and is not working, and make improvements.

Target Population

IBHS's target populations are North Carolina National Guard (NCNG) members and their families throughout the state; however, the program does not turn any service members away. All active-component and reserve-component personnel in North Carolina are eligible—regardless of duty status; service members who are up to six months past the date of their end of term of service are also eligible. Given the high rates of deployment over the past several years, there was a particular focus on deploying and returning units. There are approximately 12,000 National Guard personnel in North Carolina. The program provides one clinician for every 2,500 people in its target population.

Resources

IBHS is supported by a variety of resources, including personnel, financial, physical, and technology resources, as well as relationships with organizations within and external to NCNG. At the time of our visit in March 2014, IBHS had 12 staff members, including one state behavioral health programs director, one state behavioral health programs coordinator, two Army DPHs, one Wing DPH, four behavioral health clinicians, and three behavioral health case managers. All of the positions except the three DPHs are funded through BYR funds (the DPHs are funded through the National Guard Bureau).

All IBHS clinicians are licensed and have had their license for at least five years. They carry liability insurance of $1 million to $3 million, and IBHS keeps records of these licenses. Case managers have master's degrees in behavioral health, are typically licensed, and have done case management outside of the National Guard. IBHS avoids hiring individuals with military backgrounds because program leaders are concerned that such individuals may experience "vicarious retraumatization" when counseling

---

[1] The information in this chapter is based on interviews with program staff conducted in March 2014, along with other materials that the staff provided.

others and that they may be inclined to share battle stories. The program also hires only individuals who they know are comfortable with the uncertainties associated with being a contractor.

At the time of our visit, IBHS had received approximately $1.7 million in FY13 BYR funds. Given its year-to-year funding, IBHS plans as much as it can and then refines its plan contingent on the actual funding that it receives. The program tries to remain flexible with its staffing to accommodate the changing budget. Due to increasing budget uncertainty over the past few years, IBHS asked some of its staff to become contractors. A few of the staff left the program as a result, but the majority stayed. In terms of physical resources, IBHS leadership have offices in the Joint Force Headquarters building in Raleigh, its clinicians have offices in NCNG armories, and its case managers work from home.

IBHS uses a variety of technological resources to carry out program activities, manage its cases, track data, and increase awareness of the program. For instance, the IBHS Helpline serves as the backbone of the program and the single point of access into the program. The Helpline operates 24 hours a day, 365 days a year, and uses *voice over internet protocol* technology, which allows IBHS clinicians to receive text and email notifications when individuals call the Helpline. IBHS also uses the Veterans Health Information Systems and Technology Architecture (VistA)/Computerized Patient Record System (CPRS) as its case management system. This is the same case management system used by the VA. In addition, IBHS leverages the NCNG information technology infrastructure. For instance, the program uses the NCNG website to increase program awareness. It also has a shared drive on the NCNG network, with a shared calendar for staff and a shared library of service providers that IBHS staff can access.

Since its inception, the program has had very strong support from the highest levels of NCNG leadership. In addition, IBHS has established strong relationships with other J9 programs, such as the NCNG's EEC and Legal Assistance program (discussed later in this chapter). This has facilitated synergy across the programs and has allowed them to leverage each other's strengths. In addition to these internal NCNG relationships, IBHS has also established relationships with numerous external agencies that provide support to service members. For instance, IBHS has strong relationships with federal agencies such as the VA and the U.S. Department of Health and Human Services (especially its mobile crisis units). IBHS also has strong ties to the North Carolina Department of Health and Human Services. In particular, the program has a military point of contact at *local management entities*, which are responsible for managing, coordinating, facilitating, and monitoring the provision of mental health, developmental disabilities, and substance abuse services. IBHS has reached out to several universities and other community organizations, including Area Health Education Centers. The program tries to leverage concrete resources that will not disappear, such as TRICARE (DoD's health care system), as well as VA and other veteran centers.

Program Activities
Activities for Target Populations
IBHS carries out many activities for its target populations, including

- operating the Helpline
- developing case management support plans for individuals
- identifying resources for clients and offering referrals
- maintaining a library of well-vetted resource providers
- coaching caregivers on how to care for service members
- providing support for commanders on how to approach behavioral health issues within their units.

The IBHS Helpline serves as the single point of entry into the program, and it has been in operation continuously since it started in 2010. IBHS clinicians have a schedule of who is responsible for responding to calls to the Helpline. Clinicians usually return calls in about five minutes, and the failure-to-contact rate is almost zero. IBHS staff have business cards, but those cards only list the IBHS Helpline telephone number, not personal telephone numbers. This is done to ensure that all callers are routed through a central point of access into the program so that no one slips through the cracks.

An IBHS clinician conducts a clinical assessment with the caller and, based on that assessment, determines what level of care the caller needs. Every person who calls the Helpline is assigned both a clinician and a case manager. The clinician develops a clinical plan for the caller's care and, if necessary, can provide crisis management or stabilization, supportive counseling, and transitional support until the person can be admitted to a higher level of care. Once a situation is stabilized, the clinician works closely with the assigned case manager to develop a case management support plan. This plan provides the individual with a roadmap and the resources necessary to obtain the assistance he or she needs. After determining the support plan, the case manager connects the individual with the appropriate resources, both internal and external to NCNG. Case managers then follow up with the individual and provide safety checks to ensure he or she is safe. To facilitate consistency across its staff when carrying out these activities, IBHS has established a detailed SOP manual that describes the program's activities and staff responsibilities.

To facilitate the rapid implementation of the case management support plan, IBHS staff have developed a library of resources that are available in North Carolina. This library is continually updated, but before a new resource provider is included, program staff vet the provider to ensure some quality control over the resources to which the program refers its clients. This resource library is housed on the shared drive and is accessible by all IBHS program staff.

IBHS also provides support to both caregivers and commanders on how to deal with behavioral health issues. Anyone who is concerned about a service member can call the Helpline, and they will be coached through how to address those concerns. Increasingly, the program is assisting commanders with behavioral health issues.

Outreach

Due to licensing regulations, clinicians are prohibited from reaching out to service members directly (i.e., they cannot "shop" for patients); however, IBHS uses a variety of methods to raise awareness about the program and reach service members who may need the program's services. For instance, clinicians are housed in armories around the state, which allows them to interact with service members regularly. IBHS staff also regularly attend outreach events, including YRRP events, drill weekends, and Soldier Readiness Processing (SRP) events. In addition, IBHS relies on referrals from other organizations and on word of mouth. Lastly, IBHS reaches out to service members through command-level support and policy (e.g., operational orders and memoranda) that increases awareness.

The program has found that early outreach to deployed units can be effective. Before deployed units return home, IBHS staff travel to demobilization sites to talk with service members about the program. This provides an opportunity for the service members to sign a consent form that allows IBHS staff to reach out to them after they arrive home. The first 30 days after a service member returns home from deployment are critical, and this allows IBHS staff to check up on service members during that time.

In addition to reaching out to service members, IBHS staff provide outreach to commanders through training videos and Critical Incident Stress debriefings.

Measurement and Evaluation

IBHS has a data collection system that allows it to analyze its activities, who is using its services, and what kinds of impacts the program is having. IBHS generates daily activity reports, as well as monthly, yearly, year-to-date, and inception-to-date reports. These reports track various output measures, including

- numbers of various calls
- actions taken by clinicians and case managers
- stabilization of crises
- hours spent in direct clinical contact with clients
- hours spent researching resources, professional consultations, and case management
- hours spent completing documentation
- hours spent on outreach
- percentage of clients referred to various types of clinical and nonclinical resources
- outcomes of cases (e.g., whether a client is under the care of a therapist that he or she has been referred to).

IBHS uses the data it collects to drive program improvements and refinements. It analyzes trends in its Helpline calls to inform where it and other NCNG and state agencies should focus their support efforts (e.g., employment, financial issues, suicide prevention). In fact, it was this sort of trend analysis that revealed that employment and financial issues were the main drivers of behavior health issues among callers to the Helpline. In response, NCNG established EEC and increased resources to its Legal Assistance program to address those main drivers of behavior health issues.

In addition, IBHS asks its clients for feedback on the resources they use. The program then uses this information to inform whether it will use those resources again in the future. Lastly, IBHS monitors clinician and case manager caseloads and adjusts them as necessary to prevent burnout among its staff.

Results

Outputs

In addition to the aforementioned output measures included in the activity reports, other key output measures include the customized and prioritized case management support plans developed by IBHS case managers, the therapy plans developed by IBHS clinicians, the relationship between the client and the clinician, the relationship between the client and the case manager, and how many resources are in the resource library that IBHS developed.

Since its inception to the time of our site visit, the IBHS Helpline had received 3,036 calls—240 of those calls have required immediate intervention, hospitalization, or imprisonment. IBHS staff had also conducted 1,224 clinical assessments, 1,285 consultations with others (e.g., commanders, battle buddies), 183 family member consultations, and 102 information and referral consultations. Case managers usually have a full caseload of 25 cases.

Outcomes

IBHS's main outcomes are stabilized crisis situations, increased awareness and use of appropriate resources, increased use of benefits among service members, and increased command awareness of behavioral health issues and the IBHS program. Another outcome is the development of policy (both internally through SOPs and the adjutant general, as well as externally through the state and federal Departments of Health and Human Services) that increases awareness of IBHS and behavior health issues in general.

Impacts

The main long-term desired impacts of IBHS include increased readiness, a reduction in the suicide rate, and a change in the way the military views behavioral health. IBHS hopes to improve readiness by improving the behavioral health of the force, thereby allowing service members to focus on the mission. In addition, the program hopes to reduce the suicide rate among service members in the state over the long term. North Carolina's suicide rate has remained fairly constant over the past several years, with four to five suicides a year in NCNG (typically half of those service members have never deployed). The program understands that many factors can affect the suicide rate, but the hope is that one of the long-term impacts of IBHS will be that the suicide rate falls in part due to the program's activities. Lastly, IBHS hopes to change the way the military views behavioral health. Program staff indicated that IBHS is already having this type of impact—based on IBHS being used by all ranks—and that the perception is that the stigma associated with behavioral health issues has decreased.

Facilitators and Challenges
Facilitators
Our assessment of IBHS identified several facilitators, including strong command support (especially from the adjutant general), access to programs within J9 (as well as external referrals), memoranda of understanding with external partners, low turnover among its staff, continual information technology support from NCNG, and use of an established case management system (VistA/CPRS).

The strong command support that IBHS has received since its inception has facilitated the program's growth over time, as well as its acceptance as the primary behavioral health resource for service members in North Carolina. This command support from the highest levels of NCNG has also sent a strong message to individual unit commanders that the program is important and that they are expected to disseminate information about it. In addition, by providing access for referrals to J9 programs, as well as to external partners, IBHS can provide a broad range of support services to its clients and improve its effectiveness. The program has also signed memoranda of understanding with several external partners, which facilitates a more efficient relationship with those partners by establishing the terms of their relationship.

Low turnover among its staff has facilitated stability in IBHS and allowed the program to develop a strong team that works well together. Another key facilitator is continual information technology support from NCNG. This is especially critical for the Helpline, for which the program cannot afford to have service interruptions. By using the established VistA/CPRS as its case management system, IBHS has been able to improve its data tracking.

Challenges
Our review of IBHS also identified several challenges, including lack of response from partner organizations to which cases are referred, stress on IBHS staff, a steep learning curve for staff to learn military culture, variations across organizations that IBHS works with that make data tracking and reporting difficult, and high turnover in the DPH position. IBHS staff indicated that there are times when they make referrals to external partners, but those partners never respond to the IBHS client who was referred. To prevent these situations, IBHS staff continually solicit feedback from their clients, and based on that feedback, the program refines its lists of external partners.

IBHS staff also indicated that stress can be a challenge, especially given the gravity of the situations that sometimes arise (e.g., they receive a call from someone who is suicidal). IBHS leadership is very conscious to provide psychological support to the staff to prevent them from burning out due to stress. The program also fosters an open culture in which staff are encouraged to talk about difficult cases and strategies to address them.

Over the years, IBHS has also realized that it may take a long time for civilians to initially learn the nuances of military culture. To address this challenge, the program has developed a four-month standardized onboarding process for new hires, which helps them acculturate to the military. Program leadership feels that it would rather acculturate civilians to the military rather than hire staff with military backgrounds.

In terms of its data collection efforts, IBHS sometimes has difficulty integrating data from other organizations with its own data. This is because there are often differences in the type of data that organizations collect, as well as how and when they collect those data. This is especially challenging for a program such as IBHS that refers its clients to so many different types of service providers. To mitigate this problem as much as it can, the program relies on case managers to follow up with clients, check on their status, and update the case log as necessary to ensure accurate and up-to-date information on cases.

IBHS staff also indicated that high turnover in the DPH positions has caused some difficulties because tasks need to be reassigned when that turnover occurs. Since its inception in 2010, IBHS has had seven DPHs. Staff indicated that the root of the problem is that the DPH title is misleading. The title sounds more prestigious than the position actually is, so people use it as a stepping stone to higher-level positions very quickly. Because the DPH position is funded by the National Guard Bureau, it is unclear how much influence the program might have on changing the title of this position.

RAND's Assessment

Whether Stated Goals Are Met

We conclude that IBHS is meeting its stated goals to assess service members and their families for immediate behavioral health needs, offer therapeutic support, provide case management services, and provide referrals to federal, state, and local resources. The program has developed a robust data collection system, and it routinely uses the data that it collects to improve the program and refine its SOPs. IBHS has become so successful that it has become the primary behavioral health resource for service members in North Carolina.

Promising Practices

Our assessment of IBHS identified several promising practices that may be helpful to other state programs, including

- having a single point of entry into the program
- developing a manual of SOPs
- using an established case management system
- hiring highly trained and experienced people who will take clients by the hand and get them where they need to be
- stressing early outreach to the units
- embedding clinical staff in the armories and wings statewide
- developing relationships with internal NCNG programs and other external organizations.

By providing a single point of entry into the program, IBHS is able to better track its activities and clients and respond more quickly. This also minimizes the risk of any clients falling through the cracks. The program has developed an SOP manual that lays out the program's activities and program staff's responsibilities. The manual is a living document that is updated regularly, especially in response to problems that arise. In addition, IBHS has adopted an established case management system—VistA/CPRS. Prior to adopting this system, IBHS used data on hard copies, which restricted its accessibility. The program now also uses a shared drive to share other types of data among staff.

Other promising practices focus on the types of staff to hire, how to reach out to personnel, and the importance of developing relationships with internal NCNG programs and other external organizations. According to IBHS staff, the program has been most successful when it hires forward-thinking staff who focus on high-touch strategies to guide their clients through the entire process of addressing their behavioral health issues. When addressing behavioral health needs, it is also critical that the staff are highly trained and experienced.

Another promising practice is to reach out to units while they are still deployed to acquire consent for outreach so that clinicians can follow up with service members right away when they return home. IBHS also asks units to send a list of "high-risk" personnel prior to returning home so that clinicians can be sure to follow up with those individuals quickly when they return.

IBHS has identified that program use increased after the program began to embed its clinical staff in the armories and wings across the state. This has facilitated trust and familiarity with the clinicians.

Finally, another promising practice is to develop relationships with internal NCNG programs and other external organizations. This allows the various organizations to be force multipliers and leverage each other's strengths. In addition, it provides a broader network of resources to which the organizations can refer their clients.

Areas for Improvement

In our assessment of IBHS, we identified one main area for potential improvement: The program lacks an automated system for generating monthly and yearly reports. At the time of our visit, IBHS was entering data manually to generate these reports. This was an area for improvement that IBHS staff identified as well, and they indicated that they would like to create such an automated system.

North Carolina National Guard Education and Employment Center (EEC)

Overview
History and Goals

In 2012, the North Carolina adjutant general identified a need for increased employment support for NCNG. This need was apparent because many deployed NCNG personnel had no employment waiting for them when they returned home, which was especially problematic for those who had been employed by businesses with 100 employees or fewer. This issue was corroborated by data collected by the state's IBHS program that indicated that unemployment was a significant stressor for service members and their families. At the time, the unemployment rate in North Carolina was about 7.5 percent, and the unemployment rate within NCNG was about 18 percent. In response to this need, in early 2013, NCNG began a concerted effort to identify which service members did not have jobs waiting for them upon returning from deployment and what could be done to help them. NCNG established EEC on July 1, 2013, as a formal mechanism to help service members, their families, and veterans find employment. In response to the growing need to address employment issues among NCNG members, $700,000 of North Carolina's FY13 BYR funds were reallocated to fund the new EEC.

The goals of EEC are to establish ongoing relationships with veteran-friendly employers in North Carolina and place clients in positions that help them achieve their career goals. EEC staff indicated that because EEC is so new, they are still developing long-term goals. However, they did emphasize that EEC tries to assist service members with career development—not just immediate employment.

EEC works hand-in-hand with employers across North Carolina to help find employment for its clients. While the main EEC is located in the Joint Force Headquarters building in Raleigh, EEC staff are also regionally located throughout the state and have offices in armories and readiness centers in Clyde, Gastonia, and Winterville.

During our interviews with program staff in March 2014, they indicated that there were plans in place to expand the program by adding counselors and another center in Charlotte—a major job market in the state.

Target Population

EEC's target population includes NCNG members and their families, as well as veterans from any service who reside in North Carolina. However, any service member in need in North Carolina is eligible for assistance from program. All three of the North Carolina programs discussed in this report are making concerted efforts to reach out to reservists as well.

Resources

In FY13, EEC received $700,000 in BYR funds. The program is funded through a combination of BYR appropriations and Army money. Additionally, other staff resources include personnel from the Active Guard and Reserve program and the Active Duty for Operational Support program. At the time of our visit in March 2014, EEC was staffed with 12 personnel: one director, one operations manager, one lead employment counselor, seven regional employment counselors, one special program advisor, and one marketing manager. All of those contracts were resourced by BYR appropriations except the director (who is funded through Active Guard and Reserve) and the operations manager (who is funded through Active Duty for Operational Support). All paid EEC employees are full time. Some AmeriCorps VISTA volunteers also work with EEC.

All EEC employment counselors are retired or currently serving members of the National Guard. Program management decided to hire retired or currently serving National Guard personnel because they saw that civilians did not understand the specific needs of service members. EEC leadership felt that they needed employees who really understood the ins and outs of the National Guard and the military in general, as well as the challenges that service members face. In addition, EEC tries to hire employees with backgrounds in human resources or in recruiting, training, or development. If possible, EEC tries to hire service members who are in its database of job seekers.

EEC is housed in the Joint Force Headquarters in Raleigh, and employment counselors have offices in armories around the state. To carry out its activities, the program relies on a variety of technology. For instance, EEC uses Corporate America Supports You (CASY) Military Spouse Corporate Career Network (MSCCN) as its primary employment database and case management system. EEC chose CASY-MSCCN because the system is free of charge through the National Guard Bureau. In addition, EEC uses the American Jobs for America's Heroes employment database, USA Jobs, the North Carolina Department of Labor, and other online resources, such as Indeed.com and LinkedIn, for job postings and networking. The program uses free online career aptitude tests from military.gov and the Occupational Information Network (O*NET), an online database developed by the U.S. Department of Labor. Lastly,

EEC has a shared drive that allows employment counselors to access spreadsheets that track the progress of active cases. In FY13, the initial equipment for EEC was funded through BYR appropriations. This equipment included computers and telephones.

EEC has received support from NCNG leadership and has established strong relationships with other J9 programs that it can leverage for additional support. For instance, EEC staff indicated that they work closely with IBHS staff on a variety of fronts. IBHS staff have helped EEC staff learn how to interview service members more effectively. In addition, if EEC staff encounter a service member who they think may need behavioral health support, they will refer that service member to IBHS. Because part of its focus is on identifying educational opportunities for service members, EEC also has a strong relationship with the NCNG Education Services Office, which is colocated in the same building as EEC. This allows service members to receive both education and employment support in one place. In addition, EEC has networked extensively with employers, nonprofits, community resources, and other veteran employment programs throughout the state.

Program Activities
Activities for Target Populations
EEC provides a variety of services to members of NCNG, including career assessment and employment planning; access to employment databases, such as CASY-MSCCN and NGEN; access to EEC's network of military-friendly employers; résumé preparation and review; mock interviews; and hiring events and job fairs. During our interviews with EEC staff, some indicated that these activities relate to EEC's goals by increasing employment opportunities, decreasing stressors, increasing access to resources and referrals, and increasing job search skills. However, it is unclear how the program is measuring the impact of its activities on service member stress.

One of the primary activities that EEC employment counselors engage in is career assessment and employment planning. During our interviews, EEC staff emphasized that the goal is not just to find service members an immediate job but to put them on the path to a long-term career. After service members contact EEC, they are asked to fill out an initial intake form that assesses employment needs, and then EEC employment counselors help their clients carry out employment searches. EEC counselors use a variety of methods to identify open job positions. They use the American Jobs for America's Heroes employment database, USA Jobs, the North Carolina Department of Labor, and online resources, such as Indeed.com and LinkedIn, for job postings and networking. EEC staff also indicated that the program had lost its license to access the H2H database of job postings, but sometimes EEC does get H2H job postings forwarded to it. In addition, EEC employment counselors coordinate hiring events and job fairs for EEC clients. Throughout this process, the employment counselors assess both employers and employees to try to identify a good employment match.

EEC employment counselors help service members prepare their résumés for identified job openings and then conduct mock interviews to help them prepare for job interviews. Résumé preparation and review is an iterative process. After a service member has a job interview, EEC staff often follow up with the employer and provide feedback to the job seeker. Employment counselors also refer service members to other support service in J9 and the broader community.

In addition to working with service members, EEC staff carry out internal administrative tasks. For instance, the team meets every Tuesday afternoon for professional development and a team synchronization meeting to put everyone on the same page. They usually have a webinar and then a professional development event so that they can better address challenging cases or learn how to use new tools.

Outreach

EEC conducts outreach to both employers and job seekers. In terms of employer outreach, the program requires that employment counselors meet with one new employer each week to introduce EEC and describe the benefits of hiring NCNG personnel. In addition, to build relationships with employers, EEC often sends someone in uniform to attend career fairs. Staff also reach out to employers through H2H, Indeed.com, and LinkedIn.

To reach job seekers, EEC uses social media, including its Facebook page, the J9 website, and Twitter. In addition to meeting with service members in their offices in armories around North Carolina, EEC staff attend YRRP and veterans' career events. Employment counselors also reach out to units while they are still deployed in theater to identify those service members who are at risk for unemployment when they return home.

Measurement and Evaluation

Through its case management database (CASY-MSCCN), EEC tracks such outputs as number of active cases, job placements, and new employer contacts. CASY-MSCCN also allows EEC to run a report on the initial questionnaire used to gather data during the initial intake of a new client. At the time of our visit, EEC staff indicated that it was not mandatory for a client to complete all information on the intake form, so many of the data fields were blank. Staff try to convince clients to fill out as much information as possible on the intake form.

Information about active cases is located in spreadsheets on a shared drive, which all education and employment counselors have access to. This is especially helpful if the primary employment counselor is not available and another counselor needs to access information on the case. Employment counselors can sort these spreadsheets based on the needs of the job seeker. Program staff indicated that they did struggle with keeping these spreadsheets updated after their meetings with employers. EEC is working on its data collection system to improve data tracking and quality.

At least once a week, EEC runs reports on how many times the program is contacted and how often clients become unresponsive and why. The lead employment counselor uses these weekly reports to monitor other employment counselors to ensure that they are talking to all of these cases and spending the prescribed amount of time with each case. Once a month, EEC runs a series of reports that summarize the counselors' active cases. These data are also used to monitor where to focus program resources.

During our interviews with EEC staff, they said that they were engaging in an effort to collect more-comprehensive data on clients' needs. Several staff members were planning to try to better capture previous responses to open-ended questions over the past year in a systematic manner so that they could then analyze them. EEC staff indicated that they would like to utilize these data to identify trends so that EEC can refocus its efforts as necessary. EEC staff also indicated that their data collection and analysis efforts would improve because CASY-MSCCN was going to separate North Carolina's data system from the rest of the nation, allowing EEC more flexibility in customizing portions of the database. For instance, EEC staff would like to identify service members using different location data (e.g., by installation or zip code) but cannot do so through CASY-MSCCN's existing structure.

Results

Outputs

Key EEC outputs include the weekly and monthly reports that track counselors' activities, including their caseloads, outreach efforts, and results (e.g., number of clients that interviewed for jobs and number of clients that received job offers). EEC also tracks the number of referrals made and the number of employers added to the employer database. In addition, the program organizes career fairs and tracks the number of clients that received job offers as a result of those career fairs.

The program is helping service members find jobs and expanding its employer network. In EEC's first ten months, the program had worked with 1,043 clients, and in 391 cases, clients were able to find employment and get hired. As of March 2014, EEC had developed relationships with 85 new employers. Counselors have an average caseload of between 40 and 70 cases at any given time. In May 2014, there were 440 clients actively utilizing EEC's services to find employment.

Outcomes

EEC's primary desired long-term outcomes include (1) decreasing unemployment among NCNG personnel, their families, and veterans residing in North Carolina and (2) creating a wider net of relationships with employers in the state. It is unclear how the program plans to measure its impacts on the overall unemployment rate among National Guard personnel, their families, and veterans. It is easier to measure the program's impact on individual cases than on overall state unemployment rates.

Impacts

EEC staff indicated that the program is trying to improve readiness and resilience within NCNG, and they hope to do that by increasing employment opportunities for service members and their families. We were not provided any evidence that EEC's activities have achieved improved readiness and resilience. EEC staff members indicated that another impact of their activities is the decreased use of state unemployment benefits. We did not receive any data that quantified the savings that EEC's activities may have had on state unemployment costs. It is unclear how the program plans to measure the impact of its activities on readiness, resilience, and unemployment.

Facilitators and Challenges

Facilitators

EEC staff identified two key facilitators: leadership support and relationships with other J9 programs and state, local, and community organizations. Since its inception, EEC has had very strong support from the North Carolina adjutant general and other National Guard leaders. EEC staff indicated that leadership support and buy-in for the program is critical and that, to succeed, a program must have support from the top.

Staff also emphasized the importance of establishing relationships with other J9 programs and state, local, and community organizations. EEC has strong relationships with J9 programs, including IBHS and the NCNG Education Services Office. This has allowed these programs to leverage each other's strengths and serve as force multipliers in providing services to North Carolina service members and their families. In addition to these relationships within NCNG, EEC has established strong relationships with other state, local, and community organizations, allowing the program to broaden its network of employment opportunities and other referral services.

Challenges

EEC staff members indicated that a common challenge that they face is related to managing expectations. Some EEC clients have high expectations regarding the types of civilian jobs for which they are qualified, and EEC employment counselors need to manage those expectations. Conversely, many clients do not return employment counselors' calls because they are not motivated to find jobs. Staff members indicated that they help service members develop their own career plan, but sometimes it takes a lot of consultations to get service members to develop a plan because they are unmotivated.

EEC staff noted that another major challenge they face is losing qualified staff due to instability in funding. The EEC team has lost valuable employment counselors because the program cannot offer them the permanency that they desire. A related challenge is that because employment counselors are networking with employers, they can find alternative jobs more easily. EEC leadership indicated that because the program is so small, the loss of one employee can cause major problems on a team where everyone plays a critical role.

Lastly, EEC staff indicated that they have challenges with their current data system. For instance, the system does not easily track underemployment or changes in employment status. When clients are hired but are underemployed, the program has to mark that client as hired for a month, and then underemployed. They would like to change this.

RAND's Assessment
Whether Stated Goals Are Met

We conclude that EEC is partially meeting its stated goals; however, this is primarily due to the program being so new. EEC is meeting its short-term goals of establishing ongoing relationships with veteran-friendly employers in North Carolina and helping service members find employment. However, it is unclear to what extent the program is meeting its other formal goal of placing clients in positions that help them achieve their career goals.

Promising Practices

We identified several promising practices as a result of our assessment: EEC grew directly out of an identified need and worked with other programs within J9 to get off the ground quickly, the program has established strong relationships with community resources and partners, and the program is conducting outreach to units while they are still deployed. EEC grew out of a need identified by both NCNG leaders and IBHS data to improve unemployment support to service members and their families. To get off the ground quickly to address this need, EEC has worked closely with other programs within J9 to leverage their respective strengths. EEC has also established strong relationships with community resources and partners to quickly establish a wide network of employment opportunities and referrals. In addition, the program has had success reaching units while they are still deployed to identify early those service members who may be at risk for being unemployed when they return home. These promising practices may be helpful to other programs, especially new programs that are trying to quickly establish themselves.

Areas for Improvement

In our assessment, we identified two main areas for improvement for EEC: require clients to complete all fields of data on intake forms and continue to refine data collection methods. Requiring all fields to be completed on the intake form would allow the program to collect more comprehensive data on client needs. In our interviews, program staff indicated that they are aware of these opportunities for improvement and that they are already taking steps to address them.

North Carolina National Guard Legal Assistance

Overview

History and Goals

When North Carolina initially submitted its proposal for FY13 BYR appropriations, it included funding for Family Assistance Centers. However, when there was a delay in receiving FY13 BYR funds, J9 prioritized positions. Because J9 had identified legal and employment issues as the primary issues facing service members in North Carolina, it decided to cut funding for the Family Assistance Centers. When North Carolina received its FY13 BYR funds, it transferred the funding that was initially allocated to the centers ($300,000) and reallocated that money to fund three new legal assistants in the existing NCNG Legal Assistance program.

The goal of the Legal Assistance program is to provide service members with access to legal services at reduced or no cost. Legal Assistance operates through the Office of the Staff Judge Advocate, with the support of J9. The program supports the North Carolina adjutant general's campaign plan to improve readiness, morale, discipline, and the quality of the force by assisting NCNG personnel and their families with their personal legal affairs in a timely and professional manner. The hope is that Legal Assistance will help service members with personal, civil, or legal matters, which in turn will improve their morale and readiness.

NCNG has found that legal services are expensive (a civilian attorney typically requires a $3,000–$5,000 retainer to take on a case), and as a consequence, service members have difficulty accessing the legal system. The NCNG Legal Assistance program tries to address legal issues early, or if it is a fairly simple legal procedure, the program tries to keep costs very low. It is important to note Legal Assistance staff are prohibited from directly soliciting potential clients, and they are prohibited from formally representing service members in court. However, they can provide advice and legal counseling leading up to court motions and hearings. If Legal Assistance is unable to assist an eligible client and when an attorney referral is in the best interest of the client, the program will provide a referral. The program received the Chief of Staff's Excellence Award for Legal Services in 2008 and 2012.

Target Population

The Legal Assistance program's target population is NCNG personnel and their families; however, the program also provides services to retirees, active-component personnel, and reservists. The program coordinates some with reserve programs and is seeking to expand that coordination in response to a request by the Judge Advocate General of the United States Army that all reserve-component and active-component forces work together and find more ways to coordinate. The Legal Assistance program targets service members who are younger, less educated, and have fewer life experiences, because this demographic is particularly at risk for developing legal issues.

Resources

At the time of our visit in March 2014, the Legal Assistance program had five staff members: a chief of legal assistance (who is a federal technician and judge advocate), three legal assistants (who are contractors), and one judge advocate who is on Title 32 Active Duty for Operational Support (on very short orders—approximately 45 days). All three of the legal assistant positions were funded by FY13 BYR funds. All of the program's staff are full time, and they have varied backgrounds and educational training. For instance, two staff members are attorneys licensed to practice in North Carolina, one staff member is an Army-certified paralegal, and another staff member is a paralegal with experience in the civilian world.

In addition to its staff resources, Legal Assistance has strong connections with other programs within J9, including IBHS and EEC. Legal Assistance program staff often provide advice to clients who initially contact other J9 programs but have legal issues. This relationship allows the programs to leverage their respective strengths and expertise to provide support to their clients.

The Legal Assistance program has established broad connections within the community that may provide legal advice or services to service members pro bono or for a reduced rate. This is significant because Legal Assistance is prohibited from representing service members at trial, so service members must use other legal services if they need formal legal representation in that circumstance. The program also has established relationships with law schools in North Carolina that may have an interest in providing pro bono legal assistance. Lastly, the program has connections to other attorneys through the North Carolina Bar Association and the North Carolina State Bar, through which the program has a seat at the Legal Assistance for Military Personnel Committee. This allows program staff to communicate and network with other attorneys who may be able to provide legal services to service members at low or no cost.

In addition, the Legal Assistance program relies on technology to carry out its activities and conduct outreach. For instance, the program developed an internal client management system on the internal NCNG computer network, and the program uses its Facebook page and the NCNG main website to post information about the program.

Program Activities

Activities for Target Population

Like IBHS, the Legal Assistance program has a single point of entry—a telephone helpline that is operated by program staff. The Legal Assistance staff track cases from the time callers initially contact the program. During an intake call, program staff assess a caller's eligibility for the program's services (which includes asking a series of questions to verify eligibility and taking basic information about the legal issue at hand). Within two days, someone from the program calls the client back with an assessment of what the program can and cannot do to help.

The Legal Assistance program provides legal services in a number of areas, including estate planning and deployment readiness, family law, military administrative law, personal property, debt collection, credit issues, immigration-related issues, landlord/tenant issues, and certified tax preparation. The majority of program cases deal with estate planning and family law.

- Estate planning and deployment readiness activities include preparing wills, living wills, and health care powers of attorney, and completing Family Care Plans prior to deployment. The program also counsels service members on selecting appropriate death gratuity beneficiaries and advises service members on issues involving the Servicemembers Civil Relief Act and the Uniformed Services Employment and Reemployment Rights Act (USERRA).
- Family law activities include legal counseling on divorce, child custody, and civilian child support awards; assistance with requests for civil protective orders; and consultation with civilian attorneys representing clients with qualified domestic relations orders involving military retirement pay and annuity.
- Military administrative law activities include providing counseling on adverse personnel actions, providing advice on issues regarding Medical Evaluation Boards and the Army Integrated Disability Evaluation System, and assisting in appeals of Noncommissioned Officer Evaluation Reports or Officer Evaluation Reports.
- Services related to personal property, debt collection, and credit issues include assisting clients with creditor harassment issues, processing identity theft cases, contacting creditors when necessary, and coordinating with the North Carolina Department of Justice on client consumer complaints. Legal Assistance helps service members with immigration-related legal issues, landlord/tenant issues, and tax preparation. For instance, the program assists in preparing and facilitating immigration and naturalization cases and corresponds with U.S. Citizenship and Immigration Services, embassies, and consulates.
- Landlord/tenant services include assisting service members and their families in enforcing their tenant rights under the Servicemembers Civil Relief Act and the North Carolina General Statutes, as well as reviewing residential lease agreements for service members.
- Tax preparation services include helping service members and their families prepare and file taxes. This past year, the program had mobile tax teams, as well as tax preparation sessions at the Air Wing and a tax event in Raleigh.

Outreach

Legal Assistance uses a variety of methods to reach out to service members and increase awareness of the program. For instance, the program has a Facebook page, and the NCNG Family Programs website has a link to information about Legal Assistance. The program also has general information listed on the NCNG network. In addition,

J9 programs provide outreach to service members about the various J9 programs, and they refer service members across programs as appropriate.

In addition to outreach through word of mouth, program staff attend YRRP and SRP events where they can interact with service members and commanders. For Legal Assistance's tax program, commanders disseminated announcements every day during tax season.

The program conducts outreach to the broader North Carolina legal community. For instance, it has strong ties to local law schools and has a seat at the Legal Assistance for Military Personnel Committee through the North Carolina State Bar. This allows program staff to communicate and network with other attorneys that may be able to provide services at low or no cost.

Measurement and Evaluation

When a client contacts the Legal Assistance program, the staff use an intake sheet to collect a brief description of the issue at hand, basic demographic information, and information about the caller's eligibility to use the program's services. The program does not want the intake process to take a long time, so staff ask only a few key questions so that the attorneys on the team can assess the case. The team uses an internal client tracker—Client Information System—that is managed and updated by program staff. Staff also track each service that the program provides, an estimate for how much that service should cost, and the demographic characteristics of its clients (e.g., branch of service, grade). Staff ask clients to complete satisfaction surveys but indicated that they do not receive many of those surveys back.

The program uses the demographic information it collects to track program utilization from year to year and to gauge whether it needs to refine its outreach methods to particular target demographics. The program also tries to identify trends through the information that it collects. This trend analysis is particularly useful in preventative law because if staff notice that they are seeing many cases of a particular type of problem, they can push out information to commanders that could prevent more of those legal issues from arising. The program also uses these data to demonstrate its effectiveness to senior leaders in NCNG.

Results

Outputs

The Legal Assistance program's main outputs include number of clients served, types of services provided, number of referrals made, cost estimates of services provided, and number of tax returns filed and refunds processed. The program has data systems in place to track these outputs, which are used for trend analysis and program improvement.

Since FY11, the Legal Assistance program has provided more than 6,000 legal services, expending more than 2,770 "billable" hours, with the value of services exceeding

$351,000. In the first six months of FY14, the program provided 1,185 legal services and 524 "billable" hours, with a value of services greater than $50,000. In addition, Legal Assistance helps service members file their taxes. The program estimates that it has saved service members approximately $40,000 in tax preparation fees, and its services have resulted in $260,000 in tax refunds.

Outcomes

The main outcomes of the Legal Assistance program are increased access to legal services, legal situations that are resolved or referred to other service providers, and money saved by service members on services they would otherwise have to pay for. The program collects information on all of these outcomes and hopes to increase understanding of legal issues among service members; however, it is unclear how the program plans to collect data to measure this desired outcome.

Impacts

The Legal Assistance program is having an impact by stabilizing legal situations for service members, and it is saving service members money on legal services. However, the hope is that the program's services will also contribute to the North Carolina adjutant general's campaign plan for readiness by enabling service members to be more ready for the mission because they are not stressed about legal matters. However, the program does not appear to collect any data to directly link its activities to this desired impact, and program staff did not provide evidence of an impact on readiness and morale. It may be possible to measure the effects of the program's activities on individual readiness and morale by including questions related to those issues on client feedback surveys.

Facilitators and Challenges

Facilitators

Legal Assistance staff identified several key facilitators, including leadership support, relationships with the community and other J9 programs, flexibility in staffing and contracts, attorneys on staff who are licensed in North Carolina, and varied experience and expertise among staff. Leadership support from the North Carolina adjutant general has been very helpful to the Legal Assistance program—especially in increasing program awareness. The relationships that Legal Assistance has established with other community organizations, as well as other J9 programs, has enabled all of these programs to leverage each other's strengths. It has also allowed the Legal Assistance program to establish a wide network of legal resources that staff can refer clients to if they need formal legal representation.

Program staff also indicated that having some flexibility in staffing and contracts allowed the program to address client needs more effectively. In the past, contracts had restricted the types of legal activities that some contractors could do, and that restricted the program's ability to focus staff on tasks as needed. Lastly, the program indicated that it was very helpful to have attorneys on staff who are licensed in North Carolina, because they are familiar with local statutes and regulations, and it is very helpful to have personnel on staff who have a wide range of experience and expertise.

Challenges

The Legal Assistance program faces three main challenges: budget uncertainty, recruiting and retaining attorneys, and high staff turnover among attorneys and paralegals. Given the program's year-to-year funding, the program operates under major budget uncertainty, and that causes challenges for long-term planning, as well as recruiting and retaining staff. It is particularly difficult to recruit and retain attorneys in the program because they can find higher-paying positions as civilian attorneys. Both of these issues have caused high staff turnover among attorneys and paralegals in the program.

RAND's Assessment

Whether Stated Goals Are Met

We conclude that the Legal Assistance program is meeting its stated goal of providing service members with access to legal services for a reduced cost or no cost. The program has helped thousands of service members in North Carolina resolve their legal issues. In addition, the program has saved service members hundreds of thousands of dollars in legal fees and taxes. As the program moves forward in thinking about its desired long-term impacts on readiness and morale, it will need to identify ways to measure the effects of its activities on these desired impacts.

Promising Practices

Our assessment of Legal Assistance identified several promising practices, including immediately interacting with people who call the program, having a single point of access to the program, tracking program activities, and leveraging connections with internal and external organizations. Program staff indicated that the success of their program hinges on these four promising practices. In particular, staff indicated that it is vital to have a person provide immediate help to clients who call the program (as opposed to having a website or having an answering machine take calls). Staff indicated that having clients enter the program through a single point of access facilitates faster response times and allows the program to track cases more effectively. In addition, program staff were adamant that it is critical for a program to track its activities because it allows the program to demonstrate its effectiveness to senior leaders. Finally, the Legal Assistance program has been very effective in establishing relationships with other J9 programs and community organizations, allowing the program to build a wide network of resources that it can then leverage for its clients.

Areas for Improvement

We identified three areas for improvement in our assessment of the Legal Assistance program: insufficient use of client surveys, lack of awareness that spouses are also eligible for the program, and insufficient attorney coverage in all geographic areas. All of these areas were issues where program staff also indicated that they would like to improve. The program is currently not following up on collecting the client satisfaction surveys that the program administers; however, these surveys could be a valuable tool to help identify other areas for improvement.

CHAPTER EIGHT

Oregon Joint Transition Assistance Program and Military Assistance Helpline

Bottom Line Up Front

Oregon has two BYR-funded programs—the Joint Transition Assistance Program (JTAP) and the Oregon Military Assistance Helpline (hereafter, "the Helpline"). JTAP provides a wide range of employment services, with a goal of achieving 800 job placements in 2014. It also has a longer-term goal: to improve Oregon National Guard (ORNG) readiness and retention rates by improving service members' employment and building family resilience. The Helpline's goal is to provide 24-hour access to counseling services, and a related goal is to prevent service member suicides. JTAP provides myriad forms of one-on-one assistance, including training for interviews, writing résumés, translating military experiences into civilian skills, linking service members with employers, assisting service members with school applications, and helping with VA-related issues.

Because the Helpline is contracted to a vendor and the contract is directly overseen by the same personnel who oversee JTAP, we consider the two programs jointly. Taken together, these programs have partially met their stated goals. The Helpline is providing around-the-clock counseling and assistance, and as of RAND's site visit, there had been no ORNG suicides in that year. JTAP was on track to achieve its job placement goal as well, but the program did not have evidence that its job placements have increased overall ORNG readiness and retention or family resilience. While the programs collect data, it is not clear that data drive decisionmaking. The programs also have opportunities to improve their outreach, and accordingly have shifted resources to this area. The programs' promising practices include integrating all J9 programs to facilitate effective delivery and offering child and youth programs in concert with Yellow Ribbon and other programs aimed at service members and their spouses.

Overview

History and Goals

In Oregon, the majority of BYR appropriations go to fund JTAP and the Helpline.[1] These programs are housed within the state's J9 directorate. Both programs were established with state resources in the wake of ORNG's 41st Brigade's deployments (in particular, the 41st deployed to Iraq in 2003 and suffered significant casualties). The other factor driving the creation of these programs was the lack of active-component installations in Oregon. While reserve-component personnel may have uneven access to the programs and services located on active-component installations in other states, Oregon has no active-component installations, and thus reserve-component personnel do not have convenient access to the resources typically afforded there. For this reason, the J9 office attempts to provide as many services as possible to personnel throughout the state. Indeed, JTAP counselors (as well as J9's family assistance specialists) are strategically located around the state to meet the needs of service members. And of course, service members and veterans can receive counseling by phoning the Helpline from any location.

The Helpline's goal is to provide 24-hour access to counseling services. A related goal is to prevent suicides among service members in Oregon. JTAP focuses on providing a variety of employment services, with a goal of achieving 800 job placements in 2014. JTAP is also focused on a longer-term goal: to improve ORNG readiness and retention rates by improving service members' employment and building family resilience.

These programs began as local efforts, with a focus on veterans helping veterans. The Helpline began after the 41st Brigade's 2003 deployment, at least partly due to the efforts of a civilian who had experience with and an interest in veteran affairs. J9 pulled together people from various military offices to provide support for this effort; state funding began in 2005. JTAP began in a similar fashion, based on a peer-to-peer system initially staffed by active-duty personnel who were recovering from moderate injuries. Federal funding began in 2007 and BYR funding began in 2011, but the state has continued to make substantial contributions.

The civilian who worked to establish the Helpline also held the contract to run the Helpline for several years. However, the J9 directorate was required to put the contract out for bid, and a different vendor recently won the bid. This vendor uses licensed counselors (previously, the Helpline was staffed with trained volunteers). The relationship between the Helpline and the rest of J9 is somewhat distant; the Helpline is in a different location and J9 staff understand that the contract could go to yet another

---

[1] BYR funds also pay for personnel in the child and youth programs and marketing offices; we discuss these positions in our assessment but focus on the JTAP and Helpline programs, which represent the majority of the BYR funds. Much of the information in this chapter is based on interviews with program staff conducted in January 2014, along with other materials the staff provided. Additional sources are noted in subsequent footnotes.

vendor in the future. Finally, the state recently mandated another helpline with a focus on preventing suicides among the state's veterans. J9 staff indicated to us that they view this as redundancy.

Target Population

There are about 8,600 ORNG personnel in the state, and they have about 20,000 family members. There are about 2,200 other military personnel in Oregon, with roughly 5,100 family members. There are also some 325,000 veterans. Together, these groups constitute the programs' target population.

While JTAP and the Helpline place some emphasis on serving ORNG personnel and their families, program staff make a significant effort to assist active-duty service members, their family members, and veterans. Across all the programs within J9, the overall focus is on "employment, education, and health care," with a goal of building resilience. Thus, the goals of JTAP and the Helpline are well integrated with the goals of the other J9 programs.

Resources

BYR funding for the Oregon programs was $1.4 million in FY14. The funds went to support the contract to operate the Helpline and to pay salaries and expenses for the 15 JTAP counselors, one child and youth programs counselor, and one person who works on social media and marketing for the programs. Because the vast majority of BYR funds go to support JTAP and the Helpline, we focus our analyses on these programs; we do not evaluate the extent to which the child and youth programs or the social media and marketing departments meet their goals. Both programs are housed within J9; the J9 organizational chart lists 66 paid personnel, plus two AmeriCorps VISTA volunteers. Thus, J9's human capital and experience are considerable resources; J9 also provides office space for program staff.

JTAP's most prominent unique feature is a strong connection to the other family-oriented programs in J9, which is a relatively large organization, and most programs are colocated in the same building in Salem. J9 takes responsibility for many programs in Oregon. Because the state has no active-component installations, J9 hosts programs that might otherwise exist on those installations. Also, J9 runs multiple programs in concert (for example, child and youth programs are often coordinated with Yellow Ribbon programs).

The J9 director emphasized the need to deliver services cost-effectively and to avoid duplication, which is another reason for the close coordination across programs. Consistent with this, J9 works to create a seamless, "warm hand-off" experience for service members and family members. Therefore, while the BYR-funded programs alone do not form a wraparound program, from the service member's perspective, it is possible to access a wide variety of programs from within J9, thus creating a wrap-around experience.

Other resources include the H2H database (used by JTAP counselors), as well as considerable state support for the programs. JTAP uses the H2H database to track hires, but the program also considers H2H a partner. In particular, staff emphasized that the database is useful for following up with employers and helpful in describing military skills in civilian terms. JTAP counselors have licenses that allow them to enter information into the H2H database. This relationship is closer than the relationships between H2H and some other BYR programs (for example, North Carolina's EEC staff view H2H as a tool for reaching out to employers). JTAP staff state that the partnership with H2H helps them to place service members in jobs more effectively.

Program Activities

Activities for Target Population

JTAP activities include providing one-on-one assistance for building interview skills, writing résumés, translating military experiences into civilian skills, linking service members with employers, assisting service members who wish to apply to schools, helping with VA-related issues, and following up on progress. Helpline activities include 24-hour telephone-based counseling. (A small amount of BYR funds also goes to child and youth-oriented activities and marketing activities, as described.) These activities are integrated into 60-day Yellow Ribbon events in many cases.

JTAP counselors have few specific academic requirements. All of the counselors are former military members. This is not an explicit requirement, but the program does seek to hire personnel who have a deep understanding of the issues facing service members, as well as strong personal communication skills. After hiring, JTAP counselors receive some standardized training to familiarize themselves with the resources and training materials available, such as how to use the H2H database; counselors also undergo basic training on suicide prevention skills.

Outreach

JTAP and the Helpline have some outreach activities in place, such as electronic and paper newsletters. Also, the close integration of all of the J9 programs serves as a form of outreach; staff in other programs are likely to refer service members to JTAP, and all staff members across the program work to improve awareness of the Helpline. JTAP counselors have connections with employers, which can serve as a form of outreach. During our visit, JTAP staff placed less emphasis on their relationships with employers than did staff at some other programs, but such relationships do exist. Also, JTAP counselors are working on a county-by-county basis to build partnerships. At the time of our visit, four counties had signed covenants to support veterans, and JTAP was targeting about six other counties. At the J9 director's guidance, the programs were focused on improving outreach and developing new outreach efforts. In particular, they hired staff specifically to improve outreach through multiple routes—especially social media.

Measurement and Evaluation

The Oregon BYR-funded programs collect a substantial amount of information. The Helpline collects detailed output measures, such as the number of calls, time of the day and week that each call occurred, and type of service/referral required for each.

JTAP collects information on key activities and outputs, such as the number of interviews and training sessions delivered, as well as the main output—job placements. JTAP also collects anecdotal information on service members' successes. Within the program, these measures are used to track the progress of staff. If staff members are not meeting their goals, JTAP has a process in place that includes counseling and assistance from other JTAP staff members. Staff indicated that they also look at program data to see trends in overall placements and activities, and that they could reallocate resources based on these trends. Finally, JTAP staff complete situation report forms to report suicide ideation and actions taken. Although in most cases, J9's family assistance specialists work with service members and family members in crisis, JTAP staff also are prepared to respond should a participant have such a crisis.

The BYR-funded child and youth programs report number of participants (children and parents), outreach, briefings, and basic feedback from participants annually. (Feedback comes from short surveys and focuses on program satisfaction.) No information collected by the marketing department was available.

As a group, the BYR-funded programs are not using all of the information they collect. Indeed, in many cases, program staff gather and report data mainly to meet reporting requirements. This suggests that their overall strategy does not depend heavily on this information. Indeed, staff expressed some frustrations over a lack of standard metrics, as well as a lack of conviction that the measures collected provide a complete picture of their programs' progress.

Nonetheless, most programs have software that provides reporting templates—suggesting they could use the information in a more systematic manner. In the case of the Helpline, trends are examined on a monthly basis and resources are reallocated as necessary (this could mean shifting staff to cover different days or times, for example). In particular, the staff expect to see a spike in calls when the 41st deploys next, and they will shift resources as necessary to cover these calls.

Finally, J9 staff have discussed developing a "dashboard" that represents a multidisciplinary approach to identifying risk (especially suicide risk). For example, staff explained that there is a correlation between suicide ideation and traffic tickets; additionally, those who will eventually consider suicide may first have infractions for being absent without official leave or driving under the influence. Thus, the idea is to identify and support those who are at risk. In particular, staff have suggested focusing on those who have experienced the death of a family member to make sure they receive resilience training.

Results

Outputs

The programs provided us with several different metrics; most were gathered as of January 2014. JTAP reported nearly 1,100 individual interactions with service members and family members, as well as 434 instances of career assistance. The program also provided information from a past period, September 2011 through March 2012, during which JTAP counselors reported attending nearly 900 events and providing more than 1,900 referrals. At the time of our visit, the Helpline reported 18–47 calls per month. The child and youth programs report that 40 or fewer children took part in individual programs, but programs affected 450 children during the annual Month of the Military Child, and the program briefed hundreds of people throughout the year.[2]

Outcomes

The primary outcome for JTAP is job placements. The program reported 377 hires for FY14 as of January 13, 2014. This number can be broken down into 284 JTAP hires and 93 indirect hires; the latter category includes jobs found through other resources but while service members were taking part in JTAP. In FY13, the program reported 531 hires.

JTAP staff use two figures to calculate the impact and savings from their program: For each placement, DoD saves $24,000 in unemployment compensation, and the county saves $4,500. Staff acknowledged that these figures are estimates; no citation was provided for either. JTAP staff also track the number of interviews that take place with clients, as well as the number of one-on-one training sessions and résumé reviews. According to JTAP documents, staff provided 806 one-on-one sessions and 346 résumé reviews between September 15, 2011, and March 1, 2012.

The primary outcome for the Helpline is providing 24-hour access to counseling and referrals for service members and veterans across the state. The Military Assistance Helpline served 775 service members in calendar year 2013.

Impacts

Measuring long-term impacts from these programs is not straightforward. The Helpline focuses on suicide prevention; no J9 staff expressed any other long-term impacts aside from the obvious positive of preventing suicides. JTAP expressed broad goals of increasing readiness, retention, and family resilience; impacts, then, would include improvements to these. Staff see military retention as an especially important measure and believe that their interventions increase boot camp ship rates, which is plausible. However, existing data do not allow staff to measure the relationships between JTAP and these potential impacts.

---

[2]   April has been designated as the Month of the Military Child annually since 1986. Commands and military-affected school districts hold various events to recognize military children each April.

Facilitators and Challenges

Facilitators

One main facilitator for JTAP is that it is colocated with and very well integrated into the other J9 programs in the state. Because there is no active-component installation in Oregon, J9 takes an especially large role in providing services to military-connected residents. The integration into J9 may result in significant cross-program efficiencies, and J9 supplies additional resources to JTAP.

Challenges

These programs find it challenging to manage different funding streams and the different contractual models that result. For example, the rest of the J9 programs do not work as closely with the Helpline as they did before a new contractor won the bid to run it. Staff had a long-term close working relationship with the previous contractor, which had established the Helpline; they are less familiar with the current contractor, which does not work closely with the rest of J9. Also, J9 personnel pointed out that the state has recently established a second veteran-focused suicide hotline, resulting in redundancies and perhaps even confusion. J9 program personnel also find reporting requirements to be a challenge because they are not always certain of what constitutes success for each program.

RAND's Assessment

Whether Stated Goals Are Met

Based on our determination, JTAP and the Helpline have partially met their stated goals. Specifically, the Helpline is meeting its primary goal of providing around-the-clock counseling and assistance. The program's other goal is to have no ORNG suicides. As of our visit, there had been no ORNG suicides in Oregon in recent months, but achieving this goal does not fall completely within the Helpline's purview.

JTAP has a stated goal of 800 job placements for the year; they were on track to achieve this as of our visit. Their other goals are to improve family resilience and ORNG retention rates. JTAP tries to improve retention rates by improving service members' and their families' education, employment, and health. JTAP staff stated that family engagement services are explicitly related to boot camp ship rates, but they did not have evidence that job placements increase overall boot camp ship rates. While overall retention is certainly measurable, it is not clear that JTAP is measuring it—or that the program is able to control for other factors likely related to retention (such as the civilian economy or the deployment cycle) in the manner that would be necessary to isolate the effect of JTAP's services. Therefore, we determine that this program is partially meeting its goals.

Promising Practices

The integration of all J9 programs appears to facilitate effective delivery of services and programs. While this may be especially appropriate in Oregon due to the lack of active-component installations in the state, the "warm hand-off" approach is likely to be beneficial in many other circumstances.

The strategy of offering child and youth programs in concert with Yellow Ribbon and other programs aimed at service members and spouses has promise. While we do not have direct measures of the effects of the strategy, child and youth program staff offered considerable anecdotal evidence that this helps increase participation. J9's practice of combining appealing child and youth programs with resilience training also seems an excellent way to attract service members, deliver programming, and use resources wisely. This practice, too, is made possible by the tight integration of programs housed within the J9 directorate.

Finally, within JTAP, the combination and integration of H2H is particularly likely to yield benefits to program participants.[3] (In some other states, employment programs were run in a less integrated manner.)

Areas for Improvement

While these programs collect quite a bit of data, they do not utilize all of the data to make decisions. J9 staff are aware of this. Outreach is another pertinent area for improvement. J9 has shifted resources into this area; in particular, J9 has recently hired someone to work on social media and outreach for the programs. As is the case of many programs, these programs struggle with the uncertainty of the BYR funding stream, as well as with managing the different contracts inherent in this funding and staffing model.

---

[3] H2H generally allows one person in each state to access the database. This was a stumbling block to having JTAP counselors use H2H; however, all JTAP personnel in Oregon now have licenses to access the database.

CHAPTER NINE

Tennessee National Guard Employment Enhancement Program

Bottom Line Up Front

The Tennessee National Guard Employment Enhancement Program (NGEEP) is among the newest of the BYR-funded programs; at the time of RAND's visit, it had been operating for about four months. The primary goals of NGEEP include improving employment outcomes for service members and their families, enhancing readiness, and improving resilience. Activities in support of these goals include training for interviews, drafting résumés and cover letters, teaching service members to maintain a positive social media presence, and assisting service members in translating their military experience to apply to civilian employment. Program staff also spend a significant amount of time working with employers to match service members to appropriate jobs and to create new employment opportunities.

RAND concludes that the program's goals are partially met. The program has clearly placed service members in jobs, but it does not have the metrics in place to measure an effect on readiness or resilience—something that would be quite challenging, given NGEEP's limited scope. Accordingly, an area for improvement for this program is to develop goals that are more concrete and readily measured—for example, to focus more on the quality of the jobs that service members acquire. In addition, although program staff advise service members on using social media to their professional advantage, the program itself makes very limited use of social media for marketing and outreach. Although NGEEP is new, its model of hiring program staff to handle geographic regions and giving them discretion to tailor their programs based on local employment conditions holds promise. In addition, the program collects information on the quality of the job, and this more-nuanced information on job opportunities constitutes a promising practice.

Overview

History and Goals

NGEEP is a recently established program funded by BYR appropriations.[1] During our visit on March 20, 2014, the program had been operating for approximately four months. While it initially received funding in the early fall of 2013 and hired staff in October of that year, the government furlough delayed the initial program start. However, the program became fully operational by mid-November 2013. NGEEP is part of the state J9. Employment counselors are located around the state, but the J9 office and the lead counselor are located in Nashville.

NGEEP focuses on helping service members find employment (or better employment) by assigning each participating service member to an NGEEP counselor; assignments are based on geography. Counselors work with service members to improve a variety of job-seeking skills, including writing résumés, interviewing, and maintaining a professional online presence. Counselors also spend a substantial amount of their time working with employers in their geographic area to create and improve opportunities for service members, as well as to match service members to appropriate jobs.

NGEEP works closely with the Job Connection Education Program (JCEP)—a National Guard employment program that was established prior to NGEEP and has offices in four states. In their general approaches, the programs are similar, but they cover different areas of Tennessee. JCEP covers most of the major metropolitan areas, and NGEEP covers the rural areas, as well as Nashville. Essentially, NGEEP complements JCEP by extending support into rural areas.

Employment counselors have considerable freedom to determine how best to place participants with employers in their regions. Indeed, the clear expectation is that counselors will tailor their programs based on characteristics of the employers and participants in their regions. The primary goals of NGEEP include improving employment statistics among service members and families in Tennessee (in particular, lowering unemployment), enhancing readiness and improving resilience, and developing sustainable financial support for the program. Longer-term goals include improving the overall economic condition of Tennessee by placing service members in better-paying jobs, as well as using service members' skills to attract employers with high-skill, well-paying jobs to the state.

Target Population

Populations served by NGEEP include the Tennessee National Guard (roughly 15,000 service members), reservists, veterans, and the family members of each of these groups. Tennessee National Guard personnel are NGEEP's primary target population.

---

[1]   Much of the information in this chapter is based on interviews with program staff conducted in December 2013 and March 2014, along with other materials the staff provided. Additional sources are noted in subsequent footnotes.

Resources

This program received $500,000 to fund the initial year of operations; these funds were received in fall 2013. At the time of our visit, the program was entirely funded by BYR. Key resources include five NGEEP employment counselors, plus JCEP and other J9 programs. J9 and armories around the state provide office space to NGEEP employment counselors. Numerous organizations outside J9 also serve as resources, such as the Tennessee VA, the Tennessee Cities Military Affairs Council, and the local branch of the Wounded Warrior Project. These organizations provide employer contacts to NGEEP and help publicize the program. Finally, existing job-related databases serve as a resource, including O*NET (which includes occupational requirements), H2H, JCEP's database, and Jobs4NT.gov.

Program Activities

Activities for Target Population

NGEEP counselors provide a wide variety of resources to service members who are searching for jobs, including those searching for better jobs. Specific activities include building interview skills, holding mock interviews, working on résumés and cover letters, providing general career counseling, working with service members on how to present themselves to employers and how to maintain a positive social media presence, assisting service members in translating their military experience into terms a civilian employer can understand, and providing employment workshops. NGEEP counselors also advise service members on how to become entrepreneurs, start small businesses, and obtain small business loans, as appropriate.

In addition, counselors are responsible for identifying and coordinating employment needs of service members and family members, providing outreach to service members and employers, maintaining their database tracking activities and outcomes, participating in YRRP events, and providing unemployment statistics for tracking purposes. Counselors refer service members and family members to a variety of other resources as appropriate (for example, financial counseling, housing assistance, and behavioral health services). Finally, counselors spend a significant portion of their time interacting with employers. These activities include building and maintaining networks with employers and other organizations (such as chambers of commerce), educating employers on the advantages of hiring service members, and following up after service members interview with employers.

Counselors' caseloads vary, but each counselor is generally responsible for working with between 50 and 80 service members at any point in time. The lead counselor has significant administrative duties in addition to working with service members; therefore, his caseload is lower—roughly 20 people. Service members are considered "active" from the time they enter the program until they find a job. This system appears manageable during these early months of the program, but over time, it has the potential to create large, unmanageable caseloads of difficult-to-place service members.

Outreach

In addition to interacting with employers and organizations like chambers of commerce, NGEEP counselors use a wide variety of methods to reach both the target population of service-connected individuals and businesses that may be interested in hiring program participants. To make sure service members know about NGEEP, counselors attend reserve-component drills; this has been facilitated by supportive readiness NCOs and commanders. NGEEP counselors attend demobilization and YRRP events, as well as outreach events at the armories. Counselors also receive referrals from family assistance specialists within J9.

NGEEP does not have a social media strategy at this point. However, counselors use their personal LinkedIn and Facebook accounts to connect with service personnel. The program is considering various social media strategies as it moves forward. However, it does have a database of more than 2,600 email contacts and more than 1,000 telephone contacts among service members (these contacts are not mutually exclusive).

Measurement and Evaluation

NGEEP counselors keep track of many outputs. A primary output is contact with service members, which could be via email, telephone, or in person; counselors are responsible for contacting each of their assigned participants at least once per week. Counselors also keep records of résumé building sessions, mock interviews, and referrals, as well as all contacts they make with employers.

The primary outcome measure that NGEEP tracks is the number of job placements, but the program also tracks more-nuanced statistics related to this output. Namely, the program tracks and reports the average hourly wage of job placements and the average cost per placement, calculated by dividing the salaries and travel costs for all employment counselors by the number of placements.

Outcome measures (job placements, average hourly wage, and cost per placement) are reported only at the program level, not by NGEEP counselor. (However, the information is collected from each counselor; counselors' tools include a dashboard with contacts, their caseload, and a mechanism for reporting contacts and activities). During our visit, staff indicated that reporting these activities by counselor would prompt comparisons that they felt could be misleading due to the very different types and levels of employment available in each region of the state.

It was clear from our interviews that NGEEP employment counselors view all three metrics (job placements, hourly wages, and cost per placement) as measures of their effectiveness. In particular, the program does not explicitly compare cost per placement with other programs' costs, and there is no target cost per placement, but it is clear that decreasing the cost per placement is viewed as a measure of program success. In summary, the program tracks numerous metrics, including a potential measure of job quality (hourly wage among those who find employment through this program). The program focuses on total job placements and the related measure of cost per job placement; these measures are related to the program's overall goals. At this point, it does not appear that the program uses metrics as a means to reallocate resources.

Results

Outputs

Output measures include email and telephone contacts with participating service members, one-on-one meetings, mock interviews, résumé reviews, and referrals for interviews. The program tracks these in the aggregate. At the time of our visit, the program had recorded 2,633 email contacts, 1,080 telephone contacts, 219 one-on-one meetings, 96 résumé reviews, nine mock interviews, and 311 referrals for interviews (all counted since the inception of the program).

Outcomes

As of March 19, 2014, NGEEP had placed 137 people in jobs. This is an average of about 26 placements per month and about five placements per counselor per month. According to program statistics, however, the number of placements per month increased sharply between November and March, as the program became established. At the time of our visit, the average hourly wage of all job placements to date was $16.77 (approximately $35,000 per year if full time). The program does not track full- versus part-time status of job placements or the availability of benefits. However, the average hourly wage provides a basic measure of job quality. The total costs for program staff salaries and travel costs were about $39,000 per month. The cost per placement has decreased steadily over the life of the program (from about $3,260 to about $1,130), reflecting the dramatic increase in placements per month.

Impacts

The program's desired impacts include lowering the overall level of unemployment among Tennessee National Guard personnel, as well as increasing the National Guard's levels of readiness and resilience. Other long-term desired impacts include increasing NGEEP's visibility at the state and community levels, strengthening relationships with employers, strengthening relationships with the target population, and making sure the target population is aware of other J9 programs. Finally, this program seeks to improve the overall economic climate in Tennessee by attracting employers with high-quality jobs to the state.

Many of these potential impacts are difficult to measure. In particular, establishing causality between the program's activities and the National Guard's resilience and readiness or the state's overall economic climate is not straightforward and is beyond the scope of the program. However, NGEEP counselors are increasing visibility of this newly established program by continually making contacts with potential employers and other stakeholders, and they keep track of these contacts.

In contrast, service member unemployment is well defined and theoretically straightforward to measure. Indeed, NGEEP counselors are required by contract to measure unemployment among the Army and Air National Guard in Tennessee. To this end, the program designed a survey instrument that includes questions about educational enrollment and receipt of unemployment compensation benefits, among others. Staff administer this survey on paper at the armories. Survey results suggest that Tennessee National Guard members currently have an unemployment rate of roughly 8.5 percent. We have no information about the accuracy of this measure. Also, given the short time period that this program has been operating, there is no indication of a trend in this rate.

Facilitators and Challenges

Facilitators

Program facilitators include

- placement within the J9 directorate
- strong relationships with the other J9 programs
- ongoing coordination with JCEP
- support of veterans' issues within the state by a variety of stakeholders, including the Tennessee VA, the Tennessee Tri-Cities Military Affairs Council, and several chambers of commerce
- presence at armories
- strong military contacts (most are drilling reserve-component personnel; all have National Guard experience).

These facilitators are especially helpful as employment counselors build relationships with potential employers.

Challenges

Notable barriers include the inherent uncertainty of the BYR funding stream (which is of particular concern because this program has no other funding sources), as well as the difficulties involved in managing the employment counselors' contracts. In particular, the NGEEP director has concerns that the counselors' contracts do not allow sufficient flexibility, given that they are expected to respond to participants and attend many events outside of regular hours. He would like to be able to compensate counselors to reflect their responsiveness. While the use of flex time is helpful, it is not sufficient to solve this problem, and the director worries about burnout among the counselors as well. A related danger is inherent in the program structure—employment counselors build personal relationships with employers and tailor the program to fit their region and, quite plausibly, their own skill set. This means that counselor turnover could be very disruptive to this program and may present challenges in the future.

RAND's Assessment

Whether Stated Goals Are Met

Many of NGEEP's goals are long term in nature and difficult to measure, and the program still has a very short track record. While the program has placed participants in jobs and has collected data that suggest these are jobs of reasonable quality, the program does not have the metrics in place to measure an effect on resilience or readiness (indeed, measuring the effect of such a small-scale program on resilience or readiness poses a very challenging problem). And the program's measure of the state-level unemployment rate among National Guard personnel may be lacking in accuracy. Thus, we assess that the program has partially met its goals.

Promising Practices

One of NGEEP's most promising practices is collecting information on job quality (average hourly wage). Any additional information about job quality has the potential to help programs focus on placing service members in stable jobs that provide sufficient income. (In contrast, focusing only on number of placements could lead some counselors to push service members into unstable or low-paying jobs just to achieve a placement; measures of job quality serve to counteract this potential problem). While collecting hourly wage data is an improvement over commonly available information and constitutes a promising practice, there is the potential for this program (and other employment programs) to collect even richer data on job quality, including hours per week and the availability of benefits.

The program's model of hiring counselors to handle specific regions of the state, and giving those counselors considerable freedom also has potential as a promising practice for a very effective job placement program. Rather than following a single formula, counselors have the ability to tailor their programs based on local employment conditions and to change their focus over time if conditions change. We do note that in such a model, counselor turnover is likely to be especially disruptive if new counselors expend considerable resources trying to reinvent the wheel because there is no overarching process in place. However, with appropriate documentation and standard operating procedures that capture information about previous counselors' activities, this model has potential as a promising practice.

Areas for Improvement

Aside from cellular telephones carried by employment counselors, the program uses technology in rather limited ways. While NGEEP staff counsel participants about their social media profiles, the program itself makes very limited use of social media. For example, employment counselors sometimes use their own personal Facebook and LinkedIn pages to reach out to potential participants. The program recognizes that they need to develop a social media plan.

While the program collects considerable data on activities, outputs, and some outcomes, it is not clear that the program uses these data in a meaningful manner. In particular, the program tracks job placements—appropriately focusing on them, given its goals—but there is no plan to use data on activities and outputs to determine if resources are being well utilized or could be better allocated. This may be related to the program's short track record.

Many stated program goals are difficult to measure or tie to the program's efforts, including readiness, retention, and unemployment rate among state National Guard personnel. Additionally, measuring the state unemployment rate among guard personnel is a fairly time-consuming activity. For this reason, focusing program efforts on the more concrete goals related to job placement has the potential to improve the program's effectiveness.

CHAPTER TEN

Vermont Veterans Outreach Program

Bottom Line Up Front

The Vermont Veterans Outreach Program (VTVOP) is one of the most established programs included in this study. Its goals are to connect Vermont's veterans to the support services that they need and to enroll veterans in the federal benefits for which they are eligible. This is a wraparound program that offers highly personalized support to veterans of all services, with an emphasis on education, employment, finances, housing, and suicide prevention. Program staff conduct a needs assessment and use a combination of counseling, case management, and referrals to address clients' needs. Their support may also include house calls and transportation assistance as appropriate.

Based on RAND's assessment, VTVOP is meeting its stated goals. The program has put robust data collection systems in place that allow it to assess outcomes and carry out trend analyses to identify gaps in services for veterans in Vermont. To establish and maintain consistent quality in its outreach efforts, VTVOP has developed an extensive training program for its outreach specialists. Both of these practices hold promise for other states' programs. Although overall the program is very strong, one opportunity for improvement is to continue to refine metrics and methods of data collection.

Overview

History and Goals

VTVOP began in 2006 with funding secured by Senator Bernie Sanders.[1] In December 2006, the program received its initial $1 million congressional appropriation through the National Defense Authorization Act of 2007 to help establish the program.[2]

---

[1] Much of the information in this chapter is based on interviews with VTVOP staff conducted in February and March 2014, as well as additional materials the staff provided. Additional sources are noted in subsequent footnotes.

[2] Bernie Sanders, "Vermont Veterans Outreach Program Wins Funding," United States Senator for Vermont website, May 13, 2013; Vermont Veterans Outreach Program, "Vermont Veterans Outreach Program Brief," 2013.

Initially, Sen. Sanders intended for the program to directly reach out to Operation Enduring Freedom, Operation Iraqi Freedom, and Operation New Dawn combat veterans in their homes. During the first year, the program got off the ground with six staff members who focused on reaching out to these veterans and educating community leaders about the program. Given the highly personal approach of the program, staff went all across the state to find these combat veterans, driving 4,000–6,000 miles per month. In addition, the program initially got the word out by coordinating directly with units in the state—specifically with senior NCOs.

In 2007, VTVOP began compiling a list of veterans in the state and put many of its initial data collection systems into place. These included developing an outreach survey instrument that focused on needs assessment, as well as data entry files and mechanisms for tracking case management. The program also began formal training of its outreach specialists at this time.

In 2008, the focus of the program's mission officially changed, from serving only veterans of the aforementioned conflicts in the Reserve Component to serving Vermont veterans of all services, all components, and all conflicts. At this time, the program underwent significant growth. Three additional outreach specialists were hired, four staff were hired to field telephone calls at the new 24/7 hotline operated in the Joint Operations Center at Camp Johnson, and a program analyst was hired to compile data and assist with planning and logistics.

In 2009, the program secured its third year of funding from Congress and hired two additional outreach specialists and one additional staff member to field calls at the hotline. At this time, the program established a liaison office at the VA facility in White River Junction, Vermont. This office was tasked with serving as a conduit between the program and the VA and as a means to help Vermont veterans navigate the VA medical system. The program also hired a TRICARE representative, located at the Green Mountain Armory, to help veterans navigate the TRICARE system.

In 2010, three additional outreach specialists were hired, and four additional staff were hired to field calls for the hotline. In 2011, funding was secured directly from DoD to the National Guard Bureau, and the size of the program remained consistent.

In mid-2012, only partial funding for the program was secured. In response to this funding decrease, VTVOP reduced staff and limited its travel budget. Two outreach specialist positions were not filled, and the following positions were not renewed: six hotline operators, a media analyst, the TRICARE representative, and the program analyst. Additional funding was secured in late 2012 and then again in March 2013. At the time of our visit in March 2014, VTVOP had 20 staff members, including 11 staff trained as outreach specialists who check in with veterans in Vermont to identify potential needs and reintegration issues so that they can put them in touch with the services that they need.

VTVOP's goals are to "conduct ongoing outreach to veterans and their families; identify any potential needs, and facilitate the process to access all available services for the veterans; partner with the Family Assistance Centers, and other local, state, and federal agencies; [and] ensure the concerns of veterans and their families are responded to in a prompt and confidential manner." In addition, VTVOP aims to assist veterans of all services, all components, and all conflicts in transitioning off of Vermont state services and benefits and onto federal services and benefits (e.g., the VA benefits that they have earned).

Target Population

The initial target population of the program was National Guard members returning from combat in Iraq and Afghanistan. The program later expanded to target all combat veterans in Vermont, and it currently targets all combat and noncombat veterans of all branches of services in the state.

Resources

Staff

At the time of our visit, VTVOP had 11 outreach specialists, one VA liaison, one team leader, one outreach team leader, one program analyst, and five call center personnel. Each outreach specialist was assigned a territory, but all 14 counties in Vermont were covered by the program. All of the outreach specialists in the program were combat veterans, and one specialist focused solely on Air National Guard veterans.[3] VTVOP has found it extremely helpful to hire combat veterans who have experienced the stresses of war and reintegration issues, because they can relate to other veterans. The program also prefers to hire individuals who are recognized in their communities, knowledgeable about community support systems, and easily accessible to veterans and their families, as well as personable (e.g., easy to approach, nonjudgmental), good listeners (e.g., able to read between the lines to detect problems), and committed to Vermont veterans and their families.

To assist program staff, VTVOP has developed a set of SOPs it calls the "Battle Book." The program requires that outreach specialists undergo a host of training, including in the following areas:

- Health Insurance Portability and Accountability Act
- Vermont Agency of Human Services navigation
- VA benefit and system navigation
- critical incident stress management
- traumatic brain injury screening
- post-traumatic stress disorder detection (not diagnosis)

---

[3] The Air National Guard has unique data systems and a different deployment cycle, therefore VTVOP has found that it is most efficient to have a dedicated staff member focused on these veterans. In the program's experience, these veterans have fewer problems, and the most prevalent problems tend to be related to employment, housing, or health.

- reintegration issues
- active listening
- U.S. Army resilience
- peer-to-peer support
- military sexual trauma
- suicide prevention
- anger recognition.

In addition to personnel resources, VTVOP has technological and physical resources. For instance, the program uses the Vermont National Guard (VTNG)'s information technology infrastructure to house its internal SharePoint site, and program staff have laptops and Blackberries. VTVOP utilizes office space in VTNG buildings, including at the Joint Force Headquarters and in armories around the state. Outreach specialists use their personal vehicles for official program business.

Strong Support from Vermont Congressional Delegation

From its inception, VTVOP has experienced very strong support from the Vermont congressional delegation. However, the delegation has also held the program accountable—resulting in developing a robust data collection system and establishing metrics that allow the system to track its outcomes and impacts. VTVOP reports its outcomes to the Vermont congressional delegation monthly and briefs it regularly. In 2011, VTVOP leaders were asked to brief Congress on the program.

Strong Support from Vermont National Guard

VTVOP also received strong support from VTNG. In particular, VTNG provides office space for program staff, as well as the computer infrastructure for the program's data systems. In addition, VTVOP's 24/7 call line is housed within the state Joint Operations Center. During our interviews, some staff members indicated that this close connection had its downsides, because the program is sometimes perceived as being beholden to VTNG.

Program Activities

Activities for Target Population

Outreach specialists receive the names of combat veterans from all services, all components, and all conflicts to contact through a variety of means, including DD-214 data provided by the Vermont State Office of Veterans Affairs and deployment rosters from the Vermont Army National Guard (VTARNG) and Vermont Air National Guard (VTANG).[4] Outreach specialists also receive referrals from Family Assistance Center

---

[4] Every veteran is issued a DD-214 form, identifying the veteran's condition of discharge—honorable, general, other than honorable, dishonorable, or bad conduct.

staff, unit leadership, existing clients, family members, the VT-211 call number, and calls to the VTVOP toll-free hotline.[5] After receiving a name, outreach specialists make contact with the veteran to explain the purpose of the program, converse about how things are going, administer a survey or needs assessment, verify the veteran's VA enrollment, and refer the veteran to the appropriate help centers based on the survey answers or needs. Most outreach specialists have at least eight to ten and at most 30 active cases at any one time. A case is closed when the veteran's problems have been solved, but the case could be reopened if the veteran reaches out to VTVOP again at a later date.

One of the primary activities of VTVOP is to transport veterans to medical appointments and other support services. Because the VA is not allowed to transport service members, outreach specialists help transport veterans to the VA for their first three visits. Outreach specialists use their own personal vehicles and are reimbursed for their mileage. VTVOP explored the option of purchasing vehicles for outreach specialists to use but found that it was less cost-effective.

VTVOP also helps Vermont veterans enroll in their VA benefits and navigate the VA system. The program has a dedicated staff member who serves as a liaison to the White River Junction VA Medical Center and is located there. This VA liaison is available to assist veterans in scheduling their medical appointments, finding their way around the VA center, and filling out paperwork. If a VTVOP outreach specialist transports a veteran to the VA medical center, the veteran is transferred to the VA liaison, who can provide assistance during the visit.

Outreach

One of VTVOP's most unique features is its very personalized approach to providing support services. For instance, outreach specialists often travel to remote areas to meet with veterans living in Vermont. The program also provides veterans with transportation to medical appointments and other support services. This is especially vital in rural counties and in cases where veterans cannot drive themselves because they do not have a vehicle, or they have a medical condition or legal issues (e.g., citations for driving under the influence). In our interviews with program staff, they mentioned that these personalized efforts of "sliding our feet under the kitchen table" allowed them to help veterans more effectively.

In addition, the VA has been involved in VTVOP from the beginning of the program and continues to have a very collaborative relationship with the program. The VA provides important mental health and suicide prevention training for VTVOP outreach specialists, and VTVOP has a liaison in the White River Junction medical center who helps veterans with the VA enrollment process and navigating the VA system. An outreach specialist transports veterans to the VA for their first three visits, and when

---

[5] The VT-211 call number provides a free, confidential service to help callers find social service resources.

they arrive at the center, the liaison assists them in finding their way around the building and processing paperwork.

VTVOP outreach specialists have a multipronged approach to reaching out to a wide range of audiences, including individual units, community organizations, and other state and local agencies. For instance, outreach specialists meet with veterans in some of Vermont's 42 armories and give briefings at YRRP, SRP, and Soldier Readiness Center events. They also work with Family Readiness Groups and VTNG chaplains in the units to reach veterans.

In addition, outreach specialists connect with the broader community by networking with community organizations (e.g., veteran service organizations, churches) and sitting on external committees (e.g., mental health councils, patient-centered case committees, VA outreach committees, Vermont Military Family and Community Network). In addition, VTVOP works directly with the Vermont State Office of Veterans Affairs to identify veterans in the state, as well as with Veterans Affairs Medical Center staff to help veterans navigate the VA system. The program also works through the VA's community-based outreach clinics to identify veterans in need and connect them to services.

Measurement and Evaluation

Since its inception, VTVOP has developed a sophisticated system of data collection, in no small part because a program analyst was among the first staff to be hired. As a result, data systems were put in place early, and they have been refined over time. The program has developed its own method of managing cases—tracking the needs of veterans that are contacted, as well as the activities that are taken to assist each case.

VTVOP collects information on the needs of its clients by asking them to complete an initial needs intake survey. There are two variations of the initial intake survey: one for combat veterans and one for noncombat veterans. The program also screens for symptoms of traumatic brain injury at this time and verifies that the veteran is enrolled in his or her eligible VA benefits. Outreach specialists use the surveys and screening to assess the needs of veterans, and the information is uploaded to a case log system through a shared portal. Each case log has a unique identifying code for each veteran, and this log is a live document that is updated each time there is a new development or interaction with that individual. The program has a very flexible definition of what it considers open cases, and cases are reopened if the individual has an issue.

VTVOP carefully tracks the demographics of the veterans it contacts. For instance, it tracks the number of veterans it contacts, the number of deployments the veteran has had, the types of issues the veteran has, and the time period that each veteran received assistance. This information is broken down across branches of services and the VTNG (both VTARNG and VTANG), as well as across regions of the state. The program also tracks the percentage of personnel it has contacted in units that have deployed.

In addition to collecting information on the needs of its clients, VTVOP collects information on the program's various activities. For instance, VTVOP outreach specialists are required to complete a summary of their activities each week, including the cases worked (by type of problem, component, and activity), the briefings given, the referrals given, and the training attended.

In addition, VTVOP outreach specialists are required to complete a monthly progress report of their activities. This monthly progress report includes all meetings and events attended, all training and data tracking activities, new cases and follow-ups, monthly wellness calls, an assessment of the outreach specialists' office and workspace environment (e.g., whether the workspace is clean and orderly, whether files are protected), and current case load (e.g., number of open and closed cases, the types of issues being addressed). These data serve as the basis for the outputs that the program produces on a weekly and monthly basis to track its activities and make adjustments to its outreach as necessary. VTVOP is working on developing a Microsoft Access database to better track data.

VTVOP uses the information described here in various ways. For instance, it uses the initial needs surveys and traumatic brain injury screenings to identify the needs of veterans and develop assistance plans for them. In addition, the program uses this information to track the needs of veterans in the state and refine its outreach efforts in response to those changing needs. For instance, data collection efforts have enabled VTVOP to identify that mental health problems, unemployment and underemployment, and homelessness have become the most prevalent issues facing the veterans that have been in contact with the program.[6] As a result, the program has increased its outreach efforts and connections to providers in these areas.

In addition to utilizing the data it collects for program improvement, the data feed into the monthly update that the program provides the Vermont congressional delegation and updates to VTNG. The program also uses its data collection efforts to implement evidence-based practices. For instance, the program conducts formal case studies, selected randomly, to identify what went right and what went wrong in those cases. Those lessons are then incorporated into the program's manual of SOPs, or Battle Book. The program has found that the Battle Book is very helpful in training new staff and bringing them up to speed quickly on internal processes and expectations.

---

[6]   Program staff indicated that over the past two years, homeless young people have increasingly joined VTNG, but those that have not deployed do not have some benefits.

Results

Outputs

As indicated, the information that VTVOP collects feeds into various outputs that the program creates and allows it to better fulfill the program's mission. These outputs include the needs assessments, follow-up plans, and case logs that are generated for each veteran that the program serves. Other key outputs are the enrollment of veterans in the federal benefits for which they are eligible, the transportation of veterans to medical and other appointments that they otherwise would not have been able to attend, and the weekly and monthly reports that are compiled on the program's activities. Those reports are then analyzed and used to update the Vermont congressional delegation.

As of January 2014, VTVOP had completed needs assessments for 4,571 Vermont veterans since its inception. In addition, the program has surveyed 95.4 percent of the 1,684 VTNG members who were deployed from January 2011 to January 2014. Of the 789 veterans with multiple deployments that the program has surveyed, program staff identified that 34 percent of those service members had multiple issues. The most prevalent problem among this population was with veteran benefits, followed by financial issues. As of January 2014, VTVOP had 417 open cases. In those open cases, most were members of the VTARNG, and the most prevalent problems by a wide margin were behavioral issues (182 cases, 172 of whom were VTARNG members) and physical issues (158 cases, 146 of whom were VTARNG members).

Outcomes

In addition to these outputs, VTVOP has achieved outcomes. For instance, one of the immediate goals of the program is to provide both short-term and long-term assistance to veterans in Vermont. The data that the program collects indicate that most veterans that utilize the program only need short-term assistance (three months or less) before their issues are resolved. Of the veterans the program provided assistance to from January 2011 to January 2014, 73 percent of cases needed short-term assistance (three months or less), 14 percent needed medium-term assistance (three-to-six months), and 13 percent needed long-term assistance (six months or more). As a result of its activities, VTVOP is reducing the number of veterans in Vermont who are in need and whose needs are not addressed.

In addition, the program has successfully increased the use of federal veterans benefits among veterans in Vermont—a main goal of the program. Since 2007, VTVOP has helped approximately 7,000 veterans enroll in federal VA benefits, thus providing a cost savings to the state.

Impacts

This program's primary impact is to reduce the number of veterans who are in need and whose needs are not addressed. While the program's short-term impact is resolving acute transition issues, its long-term impact is facilitating the successful reintegration and transition to civilian life; this includes resolving long-term issues and fully using federal veterans benefits. The data that the program collects allow it to measure how many veterans it is assisting, the status of cases (open or closed), and what the issues are in each case.

Facilitators and Challenges

Facilitators

One of VTVOP's main facilitators is the strong support that it receives from the Vermont congressional delegation. Sen. Sanders has been a very strong advocate for the program and its mission, which has helped to secure initial and subsequent funding. In addition, VTVOP has had very strong support from military leaders in Vermont. In particular, VTNG has provided various resources that the program would have had to invest in on its own, including information technology infrastructure and office space in Burlington and in armories around the state.

Another key facilitator is the program's ability to collect and analyze data. This has allowed the program to identify the needs of veterans in Vermont (one of its key goals) and refine its outreach strategies to meet those needs. Because the program began collecting data soon after its inception, VTVOP is able to identify longitudinal trends and address those as well. This type of data collection and analysis has also allowed the program to provide performance metrics to both political and military leaders in the state and make a case for additional funding.

The close relationships that VTVOP has developed with the VA, as well as its good working relationships with state, local, and community partners, have been additional facilitators. These relationships have allowed the program to develop a strong network of partners to better serve the needs of veterans in Vermont. Lastly, program staff indicated that using veterans as outreach specialists makes it easier to relate to clients and garner their trust, as well as to create relationships with some community partners—especially veterans service organizations.

Challenges

We noted four main challenges for VTVOP. First, program staff noted that funding was one of their biggest challenges. In particular, they indicated that finding the right funding sources was difficult, as was the uncertainty associated with year-to-year funding. They also noted that it is sometimes difficult to find staff who are willing to live with the uncertainty of having single-year contracts. Second, program staff indicated that it is sometimes difficult to acquire accurate information on veterans living in Vermont, and this problem needs to be solved because with the drawdown, it will be even more critical to be able to track veterans in the state. Third, they noted that working in a rural state like Vermont can be challenging. In particular, it can sometimes be difficult to reach veterans in rural areas (e.g., they may not have cellular telephone or Internet service). In addition, transportation can sometimes be a challenge (especially in rural areas) and the program needs to be cautious not to become a taxi service. Fourth, some program staff noted that being connected to VTNG can sometimes make working with the other services difficult, because VTVOP is perceived to be a program focused on guard members.

RAND's Assessment

Whether Stated Goals Are Met

We conclude that VTVOP is meeting its stated goals. The program is reaching out to veterans in the state, identifying their needs, and connecting them to the services that they need. The program is also assisting veterans in enrolling in their VA benefits. Given the data collection systems it has put into place, the program is able to conduct process improvement based on evidence-based practices.

Promising Practices

VTVOP has developed several promising practices that could be helpful to other programs. For instance, over the past few years, as the funding environment has become more uncertain, VTVOP has developed a strategy to phase its contracts so that contracts for the most essential personnel end last. This has allowed the director to develop contingency plans that prioritize tasks and services for as long as possible if funding is not renewed. In addition, VTVOP has established a dedicated liaison to the VA to help veterans navigate the VA system. The VA liaison appears to be a successful means to help veterans work their work into and through the VA system. In addition, program staff were adamant that without guaranteeing confidentiality to their clients, VTVOP would not be able to acquire the trust of veterans and assist them.

Another set of promising practices focuses on data tracking and process improvement. Including a program analyst on staff seems to have played a pivotal role in the early establishment of data systems and metrics. Having a dedicated program analyst has facilitated the development of data collection and tracking systems. The program has improved its systems and refined its metrics over time. This has, in turn, facilitated quality improvement. For instance, VTVOP has been able to use the data that it has collected to focus on the areas of most need. In addition to making data improvements, the program has engaged in process improvement in other areas, such as establishing a manual for SOPs (the Battle Book) and continually updating it based on lessons learned.

Areas for Improvement

Our assessment indicates that a main area for improvement for the program is to leverage opportunities to refine metrics and methods of data collection. Our interviews with program staff indicate that they are aware of this opportunity for improvement and that they are willing to reassess their data systems. An indication of this is that staff are working on developing an Access database to better track data.

Program staff indicated that they would like to establish even deeper collaborations with some community organizations. This would allow them to provide more seamless support for veterans in Vermont, across a wide range of support services. The program is already taking proactive steps to deepen those relationships by increasing coordination with community organizations.

CHAPTER ELEVEN

Washington Employment Enhancement Program

Bottom Line Up Front

Washington's Employment Enhancement Program (EEP) aims to produce employment-ready service members within 30 days, to change the way they seek employment, and, ultimately, to place them in jobs. EEP focuses on employment-related activities that include occupational assessments, résumé and interview preparation, mock interview sessions, job search counseling, and referrals to job resources. Program staff called attention to the personalized nature of the program and described it as being "high touch" instead of "high tech." In other words, the program emphasizes personal, intensive support of clients throughout the employment search process. EEP also conducts outreach to a variety of organizations in the community (e.g., local businesses and colleges, local VA and Department of Labor affiliates).

The program's limited tracking of outputs and outcomes makes it difficult to assess whether it has met all its stated goals. This is clearly an opportunity for improvement; although some templates are available for tracking outputs and outcomes, tracking and analysis happen only sporadically. However, available data indicate that the program has met at least some of its goals: Hundreds of clients have found employment after receiving support from EEP staff. The program's emphasis on high-touch support and its focus on empowering clients by improving their employment search skills are promising practices that may be transferrable to other states.

Overview

History and Goals

The BYR-funded program has its roots in efforts to provide employment support to Washington National Guard (WANG) personnel returning from deployments.[1] Tom Riggs, a veteran and later director of WANG's Joint Services Support (JSS), worked

---

[1] Much of the information in this chapter is based on interviews with program and WANG staff conducted in January 2014, along with written responses and other materials the staff provided. Additional sources are noted in subsequent footnotes.

with a temporary technician in 2006–2007 to help guard personnel with résumés, job applications, and employment searches. By 2008, Riggs had developed a partnership with the United Association of Plumbers, Pipefitters, Steamfitters, and Piping Specialists of the U.S. and Canada that featured an apprenticeship program for guard personnel (18 weeks of a basic pre-apprentice training course, with equipment, plus the apprenticeship for 16 students). He and others also developed "Project 100," in which they sought ten employers who would each commit to hiring ten guard personnel. Building on these successful initiatives, Riggs and others pitched a program idea to John McWilliam, deputy assistant secretary of the Veterans' Employment and Training Service, Department of Labor, that was intended to augment the department's Local Veterans Employment Representative (LVER) Program staff. McWilliam agreed to support the program, and the initial employment transition coaches (ETCs) were trained alongside the LVERs and Disabled Veteran Opportunity Program specialists. In July 2009, ETCs began delivering employment assistance to demobilizing members of the 81st heavy brigade combat team while they were still on Title 10 active duty and located at Fort McCoy, Wisconsin. In 2011, EEP received its initial round of BYR funding, and, at the time of our study, ETCs were supporting service members, veterans, and their dependents from four locations across Washington.

In response to our inquiry about what EEP's goals and objectives were, program staff responded, "The Employment Enhancement Program works to reduce unemployment within the Washington National Guard by providing 'coaching' to those seeking assistance with employment." Through our interviews, we learned the goals were a little more nuanced and included producing employment-ready service members within 30 days and changing the ways service members search for employment. These were viewed as specific, intermediate goals that contributed to the ultimate goals of the program: placing service members in jobs and reducing the WANG unemployment rate.

Target Population

There are roughly 20,000 guard and reserve personnel in Washington state, with an unemployment rate of 7.8 percent, which informs the target population size. EEP staff maintained that it is difficult to estimate unemployment figures for WANG, however, because some guard personnel drop out of the labor market, some do not apply for unemployment, and some do not state that they are part of WANG when they apply for unemployment compensation. The program aims to help underemployed guard personnel as well, but this group's numbers are even more difficult to estimate.

Although the program is administered at WANG locations (e.g., Camp Murray in Tacoma) and falls under the oversight of the WANG JSS director, EEP services are very inclusive; they are advertised as open to soldiers, airmen, veterans, and their families. During our site visit, we heard about cases in which active-duty personnel based at

Joint Base Lewis-McChord (JBLM) received employment coaching. The focus, however, appears to be on guard personnel.

Resources

Since the ETC program started, it has been supported by various funding sources and, in response to funding changes, has evolved in size and scope. ETCs were first hired as temporary technicians using Overseas Contingency Operations funds. In February 2010, after those funds went away, Riggs and others created the National Guard Employment Enhancement Project (an earlier version of EEP) and made the first request to the National Guard Bureau and the congressional delegation for support. EEP has received BYR funding for three years: $1,400,000 in FY11, $467,000 in FY12, and $590,000 in FY13. In FY13, the program was wholly funded by the BYR appropriation, which supported six ETCs (down from a program high of 16) and one social media coordinator, all contractors. One ETC is located in Spokane, one in eastern Washington, one in Kent, and three in the Seattle/Tacoma area. Half of the ETCs have military experience. Some of them have pertinent college experience, and others have human resources or case management experience. All have several years of experience as ETCs and worked in YRRP or other JSS programs.

EEP draws on resources available for free or at limited cost. WANG provides program staff with a dedicated location at Camp Murray, and ETCs have facilities as needed at armories and drill locations across the state. WANG also furnishes computers, telephones, and other office equipment, and it is through WANG that the program maintains a dedicated website. The social media coordinator has taken advantage of free social media options, most notably Facebook, LinkedIn, and Twitter. Businesses and schools in the community are also a source of support. For instance, EEP has been granted free use of the World of Work Inventory (WOWI) instrument, which ETCs found especially helpful for clients' initial occupational assessment, and the Tegrity online résumé tool was made available to clients for free by a local technical college with which EEP has partnered.

More generally, the program's partnerships and other interactions with individuals and organizations serve as a force multiplier. ETCs work closely with program staff from H2H, ESGR, and YRRP. Family assistance coordinators (FACs) and transition assistance advisors (TAAs, who are staff for other guard programs) serve as ambassadors for EEP, helping with outreach to prospective clients. Looking beyond WANG programs, members of the Society for Human Resource Management shared their expertise with the ETCs, advising them on such topics as online résumés and job applications. Bates Technical College instructors provided "Tomorrow for Today" training, and the Washington Department of Veterans Affairs is an EEP training partner. Local unions and employers serve as a resource in various ways, including job training and job opportunities.

Program Activities

Activities for Target Population

The ETCs are the heart of the program. During our site visit, ETCs outlined the following activities, which were all focused directly on the client. Some activities are in-person, others are combined in-person with telephone or email follow-up support, and still others are only provided remotely (e.g., when a client is not located near an ETC).

- Job/occupational assessments: These may include a combination of interview, WOWI assessment, H2H assessment, and the matching of military skills to civilian jobs using O*NET, a website sponsored by the Department of Labor.
- Résumé preparation and support: The nature of this service varies from client to client; it may include one-on-one coaching, the use of the Tegrity online résumé site, classes, or the use of templates. ETCs may assist with building résumés for federal jobs or private-sector civilian employment. The goal, though, is not to create the résumé while the client passively observes but rather to teach him or her to create a résumé and tailor it for different jobs.
- Interview techniques: This includes "do's and don'ts" guidance for interviews and interview preparations tips, including advice provided by Society for Human Resource Management members.
- Mock interviewing: This service is an in-person interview practice with a panel of ETCs or other JSS staff. The panel asks the client typical job interview questions and then provides feedback for improving the client's interviewing style.
- Networking connections: Clients are taught how to set up social media accounts, such as Facebook and Twitter, as well as how to employ them for professional (versus personal) use. In addition, there are occasional networking events with employers, which at the time of our study were hosted at JBLM.
- Dissemination of job opportunities: Although it is not intended to serve as a job bank, EEP does receive job announcements, including information directly from local businesses with which EEP has an ongoing relationship. Sometimes ETCs will match people to opportunities; other times they will use social media for broader dissemination.
- Support for online job searches: This includes helping to repackage a résumé for submission on online job application forms.
- Material support: This typically takes the form of referrals to and help with applying for resources of this nature (e.g., unemployment compensation, food pantry).
- Limited ongoing apprenticeship programs: This has waned since the original apprenticeship programs in 2008, but there are still opportunities with Veterans in Construction and Electrical and Veterans in Piping.

Outreach

EEP outreach is directed to many different audiences and takes many forms. Outreach includes reaching out not only to those within WANG, such as guard personnel, commanders, and guard program staff, but also to JBLM-based staff, employers, HR professionals, university personnel, and more in the local community. In this way, information about the program is broadly disseminated for multiple purposes. Specifically, these efforts not only educate prospective clients themselves but also lead to more prospective clients, sources of jobs for clients, and resources that could help clients.

EEP uses many social media resources, including a website, Facebook, LinkedIn, Twitter, and YouTube. EEP is listed in the WANG J1 directory and produces various types of informational brochures, newsletters, and flyers, some of which were shared with us. The program ensures that it has some involvement with job fair events hosted at JBLM and that someone represents it at local YRRP events—for example, a FAC or TAA will talk up the program while there. Finally, FACs bring flyers with them to drill weekends:

Measurement and Evaluation

The nature and extent of efforts to collect indicators of EEP usage and effectiveness have varied since the program's inception. In 2009–2010, before the receipt of BYR funding, client tracking was neither established nor standardized; it varied by ETC. By 2011, tracking indicators and processes were emerging (e.g., use of a self-created Access database), and in 2012, guidelines were established to hold individual ETCs accountable for caseload tracking. These guidelines were in response to a February 2012 EEP audit by the U.S. Property and Fiscal Office as part of a larger audit of YRRP. The focus was on fiscal law, with an emphasis on contract structure and compliance. The 2012 Audit/Inventory Recap that we reviewed indicated that there were problems with keeping track of activities: Some client files were not in the database, some database entries did not have a paper client file, and some files had inadequate information. During our interviews, we learned that at the time of the audit, some actions were documented only within a narrative, making tracking across clients difficult.

Accordingly, the auditors' recommendations included a required minimum level of documentation in each client's paper file, employment figures reported accurately to reflect services provided and clients supported, and both ongoing surveillance of contractors and reviews of deliverables to ensure the services provided were consistent with the quality assurance surveillance plan. Templates were also made available to provide ETCs with ways to systematically document other activities, including a partnership event/visit report and a visit checklist for armories (with questions like "Are the hours and contact info for the ETC posted in plain sight?" and "Does the Readiness NCO at the armories served by the ETC have a basic understanding of the ETC?"). We did not see evidence of these materials' use, however.

In 2013, the NGEN database was adapted for EEP use, and there is a form now in place for collecting the following outputs on a weekly basis: walk-ins, new intakes, cases worked, final résumés, Tegrity use, skills training, H2H hires, apprenticeships, and close-outs. Outreach output metrics listed on the form include briefings, number of attendees, and number of business outreach contacts/resources. Finally, the form also captures referrals to JSS and EEP partners. At the time of our site visit, the form seemed relatively new, and neither the form nor NGEN seemed to be widely used yet, perhaps due in part to their newness. Some of the information was tracked less formally; for example, one ETC discussed writing down his activities and outputs on his wall calendar.

With respect to information *usage*, measures seem to be analyzed primarily in response to congressional inquiries, the aforementioned 2012 audit, higher-level reporting requests, and other external requests for results. Measures produced in response to external requests or when seeking funding include cost savings, return on investment, and other financial indicators, but the staff we interviewed could not explain how these measures were computed. In addition, some information, such as clients hired, was shared with commanders and senior leadership in the hope that demonstrating success would encourage them to push personnel to EEP. We did not find that program staff conduct trend analyses or evaluations or otherwise use data for internal purposes (e.g., as a basis for change), at least not formally. It is possible that individual ETCs track their own progress and make changes based on that.

Results

Outputs

We did not receive any data collected via the output form described in the Measurement and Evaluation section of this chapter, and it is not clear if the form is being used consistently. Additional outreach outputs have been collected at least sporadically, including the number of fans or contacts for various social media outlets. For example, in a briefing given in the 2012–2013 time frame, the JSS Washington (JSSWA) Facebook page had 1,376 fans and the JSSWA Twitter account had 485 followers. We did not receive more-recent figures. Program staff indicated that they can collect hits on the JSSWA website, but those figures were not provided to us.

Outcomes

During our interviews, program staff identified several outcomes for EEP clients. First, in the short term, the program would like to equip clients with the tools needed to be employment ready within 30 days (e.g., savvy in the résumé and application process, basic or improved interviewing skills, good networking skills). This assumes that clients are motivated and willing to exert effort in pursuit of their employment goals. As more time passes, clients should exhibit changes in employment search practices or start engaging in searches independently for the first time. Ideally, clients would be able to weather underemployment or unemployment and persist with their job search. These outcomes are reflections of the program's goals, but they have not been measured consistently, and some of them, like résumé-writing skills or interviewing proficiency, are harder to measure. The ultimate outcome, actual employment, is a more tangible, observable outcome and is the one that the program has focused on measuring.

We received several different reports that provided measures of clients hired, making it difficult to summarize the program's accomplishments on this front. Figures from 2011–2013 monthly ETC program recap reports indicate 62 clients hired in 2011 (based on three months), 238 clients hired in 2012, and 123 clients hired in 2013.

In the U.S. Department of Labor's Veterans' Employment and Training Services Request for Funding, it was reported that the program had 639 clients hired in FY10 and 406 in FY11.[2] A seemingly comparable figure for FY12, 388 new hires, was provided in an undated JSS directorate briefing. It is possible that some of the differences in numbers stem from calendar year versus fiscal year bases of measurement. These reports also included average starting wages for clients of $15.38 per hour in FY10, $18.38 per hour in FY11, and $17.08 per hour in FY12.

Impacts

EEP is intended to promote resilience for guard personnel and their families. Program staff also framed it as a "readiness enhancer": If soldiers do not have jobs and cannot provide for their families, they are going to stop coming to drill and will not be focused on the guard. There was no evidence available to demonstrate that EEP had these impacts, however.

Clients' reduced need for financial assistance and other entitlements was also identified as having state- and federal-level impacts, as follows:

- FY10
  - Federal savings: $2,090,496 (unemployment compensation savings minus operating cost)
  - State savings: $1,458,228 (based on reduced food and cash assistance)
- FY11
  - Federal savings: $4,379,584 (unemployment compensation savings minus operating cost)
  - State savings: $1,508,448 (based on reduced food and cash assistance)
- FY12
  - Federal savings: $8,164,994 total in unemployment compensation alone (net savings not provided).[3]

---

[2] U.S. Department of Labor, *Veterans' Employment and Training Services Request for Funding—Tab D: Enhanced Employment Program Return on Investment,* February 29, 2012.

[3] U.S. Department of Labor, 2012.

Facilitators and Challenges

Facilitators

On the positive side, having the ETCs physically located at Camp Murray and in or near armories elsewhere in the state has been beneficial; it facilitates a high-touch approach with guard personnel and makes it easier for ETCs to coordinate with other guard personnel who may be able to help them or serve as a referral. In addition, and especially in light of funding challenges, the partnerships that EEP staff have built within the community have been of great value to the program. Not only have the community connections led to donated services and support, such as the WOWI assessment tool and Tegrity résumé tool, but local employers also contact EEP with job opportunities for its clients.

Challenges

Our assessment of program documents and our site visit revealed several challenges for EEP. The tenuous nature of BYR funding and the breaks in between years of funding make it difficult at times for the program to implement its high-touch, personal approach. Two years in a row, the program had to stand down for several months. While this means the program now has an exit plan in place to execute when activities are reduced or terminated, such as guidance on where to direct clients, it may have discouraged some employment-seekers who were working with a specific ETC. The decline in funding over time has also led to a cut in ETCs, from a program high of 16 to six at the time of our study. Fewer ETCs in the state means that ETCs have never met some of their clients, making it harder to use some of EEP's signature high-touch techniques.

The perception that EEP was redundant to other resources (e.g., local employment programs sponsored by the Department of Labor or the VA) could present a challenge in some ways. However, as noted, the program attempts to distinguish itself as being high touch instead of high tech. Program staff also maintain that EEP's aggressive approach to networking enables it to provide an integrating function across resources. ETCs could and likely do overlap with other guard positions, namely FACs and TAAs. But FACs are regarded as a source of clients and not employment-focused. TAAs are described as a source of assistance in "navigating through the numerous benefits and entitlements in the DoD and VA system," which is also distinct from what the ETCs are doing. The LVERs and Disabled Veteran Opportunity Program specialists seem to offer similar support as the ETCs but to somewhat different populations. Program staff also maintained that ETCs have better access to the armories, are better connected to personnel, and are more likely to be called upon by commanders than these representatives or specialists. Coordination across these different resources is critical. Having a few of them in the same location likely helps (some FACs, TAAs, and ETCs are all at Camp Murray), but other coordination requires more effort.

Finally, we identified a potential challenge: sustaining the program after Riggs retires, which was pending at the time of our research. Riggs is the program's founder and was clearly a charismatic leader. He worked with LTC Don Brewer and state family programs to help with a transition, but during our interview, we heard uncertain comments, such as Riggs "really has been the guy who has gotten the money." His successors, which first include Lieutenant Colonel Brewer, will need to be able to both oversee the project internally and market it to external stakeholders who help to keep it alive.

RAND's Assessment

Whether Stated Goals Are Met

We could not assess whether EEP met all of its stated goals. While it is commendable that the program identified short-, intermediate-, and long-term goals, we did not see evidence demonstrating that clients were employment-ready within 30 days. Nor were data available to demonstrate changes in job search practices, either actual or perceived. Overall, the program collects very few measures of outputs (e.g., clients supported each week, number of outreach briefings to partners) and few measures of effectiveness in terms of either client-focused outcomes or impacts at the guard, state, or federal levels. It is questionable whether a program that is as small in scope as EEP can have a quantifiable impact on something like overall WANG readiness, so it may be difficult to ever measure fully all the outcomes and impacts the program claims to achieve.

That stated, the program has been tracking client hires, an important outcome, since its inception and has been able to demonstrate that its clients have obtained employment. For this reason, we conclude that the EEP has partially met its stated goals.

Promising Practices

EEP has two well-established practices that may be of value to other employment programs with similar goals or clients. First, the program focuses on giving tools to clients—"teaching them to fish" rather than serving as a job bank. EEP devotes a lot of resources toward developing its clients' job search capabilities, résumé-building skills, and interview techniques, and, for a small number of clients, EEP provides job skills training via apprenticeship programs.

In addition, the program has tried to create a distinct niche in the world of employment support services by focusing on being high touch instead of high tech. This can help clients, especially less motivated ones, navigate through a potentially confusing or overwhelming system of employment resources. The coaching and case management approach that ETCs use may also help with client accountability, because clients know someone cares and will follow up on them.

Finally, it was too early to tell at the time of our research, but adapting NGEN for EEP's use holds promise; it likely will promote case management transparency (i.e., it will make it easier for ETCs to work on each other's cases) and will facilitate analysis of program outputs and outcomes. Other programs may consider whether preexisting databases like NGEN or other tools can be appropriate for their own use.

Areas for Improvement

Tracking and analysis of results are spotty and lagging in improvement. This short-coming was noted to a degree in the 2012 audit, and it still is a concern of the program. Some improvements have been made—for example, NGEN makes tracking both more standardized and easier for ETCs, and templates now are available to track myriad outputs. But ETCs felt that this kind of work often meant time away from directly supporting clients, and they were thus not consistent in documenting their progress. On a related note, there is little internal tracking of promising practices (i.e., ETCs sharing and documenting successes). ETCs had a weekly call-in meeting to discuss promising practices, but we did not learn of any existing documentation. Program staff suggested that the lack of a designated program manager made it difficult to look for promising practices and areas for improvement and to track and analyze data. ETCs had a keen interest in these activities and tried to support them, but often they were neglected due to client needs. Because ETCs are responsible for maintaining "accurate records of service activities to provide measurement and impact of client-based services, outreach activities, number of referrals, and job placements achieved with 98-percent accuracy," those with responsibility for program oversight should identify ways to facilitate ETCs' documentation of client activities. For example, the ETC manual could include more documentation on ways to track progress quickly, such as how to use NGEN most effectively.

CHAPTER TWELVE

RAND Observations Across Programs

In Chapters Two through Eleven, we described the facilitators and challenges that individual programs faced, as well as some promising practices and areas for improvement. We regard facilitators and challenges as factors that the programs themselves have little control over but that influence their ability to meet their stated goals. In identifying promising practices and areas for improvement, we focused on aspects of program operations that were largely in the programs' control. In this chapter, we summarize common facilitators and challenges and call attention to the practices that we believe hold great promise and can be readily transferred to new and existing programs in other states, as well as to the most pressing areas for improvement.

Common Facilitators

We found three common facilitators across the programs: strong leadership support, integrated physical and/or organizational location, and access to free or low-cost technological resources. We discuss these common facilitators in detail below.

Strong Leadership Support

Support from military and political leadership appears to be a facilitator for many of the programs. For example, the program in Vermont has received very strong support from the Vermont congressional delegation, which helped establish the program and direct a more stable stream of federal funding than some of the other programs in our study receive. We also found that the program in California has received strong support from the Speaker of the House and the adjutant general, and this is viewed as important to ensuring continuity in funding and increasing the visibility and reach of the program. Similarly, staff in Indiana's program reported that leadership support within the Indiana National Guard has been critical to ensuring the continuity of the program, given that many of the positions have been funded by pulling together extra resources wherever they could be found.

Our interviews with program staff revealed that, in addition to support from state-level leadership, buy-in from command leadership is critical to ensuring that programs connect with service members in need of assistance. Programs in many states rely on unit leadership to encourage service members to use the program's services, so leadership buy-in is an outreach force multiplier. In several of the employment programs, the armories are expected to push data upward to program staff for two purposes: to provide direct contact information for service members in need of assistance and to facilitate the tracking of unemployment rates. Staff in these programs reported that leadership buy-in was important to ensuring these data are provided consistently. Leadership support has also been cited in the literature as a critical factor for success for programs and initiatives that target service members and veterans.[1]

Physical and/or Organizational Integration

The location of a program, both physically and organizationally, was also perceived as a factor in program success. Staff in Indiana, North Carolina, Oregon, Tennessee, and Vermont reported that the organizational placement of their programs under the J9 directorate has been useful. According to some program staff, placement under the J9 ensures a common sense of mission, integration with other programs, and support from leadership. It is not clear, however, that the programs must be located within J9 to experience these benefits. For example, Florida reported integration with other programs in J1, where its program is housed. It may be that the advantages are due at least in part to the BYR-funded program's close physical proximity to other programs that are focused on service member well-being (e.g., employment, mental and behavioral health). In Colorado, the Marketing and Outreach Program is physically located in the same office suite as all of the state family programs, which program staff noted not only encouraged brainstorming across programs but also made it easy for their colleagues to hold impromptu discussions with them about marketing needs and ideas. Avoiding appointments and travel time to different office locations also allows the program to be more responsive to day-to-day marketing needs.

Access to Free or Low-Cost Technological Resources

Several of the programs that we assessed avail themselves of technological resources that are free or have low cost, including hardware, software, and online tools. Some are used for marketing purposes while others support clients directly. For example, several states indicated that they use the computer infrastructure already in place through their state National Guard. Most of the employment programs make use of NGEN

---

[1]   Terri K. Pogoda, Irene E. Cramer, Robert A. Rosenheck, and Sandra G. Resnick, "Qualitative Analysis of Barriers to Implementation of Supported Employment in the Department of Veterans Affairs," *Psychiatric Services*, Vol. 62, No. 11, 2011; J. I. Ruzek, B. E. Karlin, and A. Zeiss, "Implementation of Evidence-Based Psychological Treatments in the Veterans Health Administration," in R. Kathryn McHugh and David H. Barlow, eds., *Dissemination and Implementation of Evidence-Based Psychological Interventions*, Oxford: Oxford University Press, 2012; Military Leadership Diversity Commission, *From Representation to Inclusion: Diversity Leadership for the 21st-Century Military*, Washington, D.C., March 2011.

and H2H software—systems intended to connect guard personnel and their families directly with employment resources, service providers, and employers. Some states, such as Colorado, have also taken advantage of free social media options, most notably Facebook, LinkedIn, and Twitter. The literature indicates that media outreach through technology has been important to programs that provide social services.[2] Moreover, Washington's EEP has enjoyed free use of both the WOWI instrument (an occupational assessment tool) and the Tegrity online résumé tool, whose use was donated by a local technical college. Other states, including Indiana, North Carolina, and Vermont, have benefited from tools created in basic software programs, including résumé templates and Microsoft Excel spreadsheets that allow counselors to track outputs and outcomes in real time.

Common Challenges

We found five common challenges across the programs: insufficient leadership support, program overlap, service members' unwillingness to fully utilize support services, funding, and staff turnover. We discuss these common challenges in detail below.

Insufficient Leadership Support

In contrast to the strong leadership support detailed earlier in this chapter, in several states, support and buy-in from leadership at certain levels has been inconsistent, leading to challenges marketing the program and obtaining data. For example, Florida and Indiana rely on command leadership to support program outreach and encourage individuals with employment issues to use the program. However, differential participation across armories suggests that some in command leadership have not made this a priority. To identify individuals to contact, Indiana attempts to collect lists of unemployed guard personnel from units. It has been a challenge for ECP to ensure that these data are provided regularly. Colorado's program has faced similar challenges with uneven use of its services, and these challenges are due at least in part to the lack of a mandate for family program offices to use the Marketing and Outreach Program.

Potential Program Overlap

In several cases, programs face challenges with confusion between the BYR program and other programs that provide similar services to veterans. For example, staff members in Oregon report that the state has opened a second veteran-focused suicide hotline, essentially replicating the services of the Military Assistance Helpline. California program staff also indicate that both service members and employers often confuse

---

2   D. R. Becker, S. R. Baker, L. Carlson, L. Flint, R. Howell, S. Lindsay, M. Moore, S. Reeder, and R. E. Drake, "Critical Strategies for Implementing Supported Employment," *Journal of Vocational Rehabilitation*, Vol. 27, No. 1, 2007.

their employment program with other veteran employment programs. This confusion and lack of awareness often hinders the ability of program staff to network effectively with program participants and employers. However, both California and Indiana report that this challenge has largely been overcome by establishing close partnerships with related organizations, eliminating duplication in services, and increasing the visibility and awareness of their programs and the unique services that they provide.

Among the programs with a focus on employment, none explicitly mentioned USERRA, and only one, Washington, mentioned a partnership with ESGR.[3] USERRA generally ensures reemployment at a service member's former civilian job upon completion of deployment service, and ESGR can provide assistance to service members in retaining their prior positions. Thus, if they are not already doing so, employment programs may find it efficient to refer some service members to ESGR for assistance in retaining their prior jobs.

Service Members' Unwillingness to Fully Utilize Support Services

Many of the BYR programs in our study encounter difficulties working with their target populations, including service members' attitudes and behaviors about the programs and the services provided, as well as their unwillingness to fully utilize available programs. The challenges related to service members' attitudes and behaviors are especially evident with the employment programs. Various programs reported that service members had little motivation to find employment and that some were not ready to seek civilian employment. Program staff in California and Indiana observed instances of service members acting unprofessionally when applying for jobs. Program staff in Indiana estimated that 80 percent of service members that were referred by NCOs never followed up when the employment counselors proactively reached out to them. Moreover, those service members that did follow up frequently expected the employment counselors to do the bulk of the work to get the service member a job.

Funding

Some of the most prominent challenges identified across sites are related to the amount, timing, uncertainty, and logistics of BYR funding. Many programs indicate that the amount of funding they receive is insufficient to meet the existing need or to implement a high-quality program. Program staff in Washington, for example, note that funding issues led to a cut in the number of ETCs, making it more difficult to implement their high-touch approach to support in certain parts of the state. Other studies reinforce our finding that temporary funding can provide challenges to successful implementation of employment and mental health programs.[4]

---

[3] While our protocol did provide an opportunity for programs to tell us about key partnerships, it did not explicitly ask about USERRA or ESGR.

[4] Demetra Smith Nightingale, Nancy M. Pindus, John Trutko, and Michael Egner, *The Implementation of the Welfare-to-Work Grants Program*, Washington D.C.: Urban Institute, 2002; Lauren Eyster, Demetra Smith

In other cases, the timing or uncertainty with which programs receive funding has created challenges. The program in Washington ceased operations twice because of breaks in funding. In Colorado, the delay with which the program received funding gave it less than a year to ramp up and spend the funds. In this case, the program spent a substantial amount of the first year planning, instead of actually enacting a program. Programs also acknowledge that uncertainty as to whether the funding will continue created various problems, including a hesitance to engage in long-term planning. Balancing the time spent developing long-term ideas versus focusing on short-term programs and "quick wins" is a challenge for many of these programs.

Finally, programs recognize problems that result from the logistics related to their funding. Certain states have had to manage multiple contracts and coordinate across multiple funding streams. In Oregon, for example, BYR funds support multiple programs that are integrated with other J9 programs. Thus, leadership working with BYR-related programs must manage and coordinate programs and staff with different contracts. Program staff in New Hampshire felt that there is ambiguity as to how their funding could be spent and what populations it could be spent on. For example, they explained that their program could not support Coast Guard members because the funding came from DoD, and the Coast Guard is part of the Department of Homeland Security. Given these funding-related challenges, most programs could benefit from seeking out and managing funding streams more strategically (a few programs have indeed begun to do this; see the section on promising practices).

Staff Turnover

Program staff often possess specialized knowledge or have developed close working relationships with employers, social service agencies, and other resource providers, which makes staff turnover especially problematic. At the same time, funding uncertainties and job-related stress may make such turnover more likely. Funding uncertainties create a lack of job security, and given the nature of BYR appropriations, many states retain program staff on yearly contracts (or shorter). For example, program staff in several states noted that the uncertainty in funding has led them to look for jobs for themselves in conjunction with helping service members find employment. Program leadership at North Carolina's EEC noted that the program lost qualified staff specifically because of funding and program instability. Furthermore, staff noted that the nature of the job is frequently stressful. For instance, program staff in Tennessee are expected to attend events and remain responsive to program participants outside of regular business hours. Consequently, program leadership in Tennessee expressed concern about staff burnout and turnover.

Nightingale, Burt S. Barnow, Carolyn T. O'Brien, John Trutko, and Daniel Kuehn, *Implementation and Early Training Outcomes of the High Growth Job Training Initiative: Final Report*, Washington D.C.: Urban Institute, 2010.

Promising Practices

As part of our assessment, we asked programs to identify practices that they felt helped them make progress toward program goals, and we also identified practices that we perceived as both useful to the programs and transferrable to other states. Where possible, we focused on practices supported in the literature. Because even programs that have not yet met their goals can engage in promising practices that help make progress toward their goals, we included all 13 programs analyzed in this report. We also turned to the literature to see if our observations and those of program staff could be corroborated. Overall, we found that programs are implementing quite a few promising practices. These include

- close alignment between activities and goals
- single point of entry
- high-touch, one-on-one services
- geographically dispersed staff
- activities that build long-term client skills
- strong partnerships with other resource providers
- strong partnerships with community organizations and employers
- early engagement with mobilized units
- broad and creative outreach strategies
- strong technology-based tracking systems
- efforts to make programs replicable and sustainable.

Close Alignment Between Activities and Goals

Ensuring that activities are closely aligned with the goals of the program is a promising practice implemented by several of the programs; this alignment ensures that resources are being devoted efficiently to achieving the outcomes that are set out for the program. California's Work for Warriors program strongly emphasizes the importance of a continued focus on achieving placements at low cost. Employment counselors are advised to focus exclusively on efforts that will lead directly to placements; for service members with nonemployment-related needs, counselors refer them to relevant organizations and programs for help. In addition, in an effort to incentivize focus on placements, Work for Warriors only tracks outputs and outcomes that it believes are closely tied to these key activities. Similarly, North Carolina's programs maintain a tight focus on activities that are directly aligned with their goals. Program staff indicate that with the IBHS program focusing primarily on behavioral health issues, Legal Assistance focusing solely on legal issues, and EEC focusing primarily on education and employment issues, these programs are able to establish clearer metrics that are tied to each program's goals, and they can still refer service members across programs as needed. By analyzing the data that it collects, Vermont's program has been quite successful at refining its activities and metrics over time to more closely align with its primary goal of reaching out to Vermont veterans in need.

Single Point of Entry

Having a single point of entry for service members to connect with the program is a promising practice for ensuring that service members are linked to services seamlessly and are continuously tracked as they receive services. For example, the only way for service members to enter North Carolina's IBHS program is through its Helpline, and counselors are encouraged to use business cards that include only the Helpline number (as opposed to individual contact information). The Helpline is heavily advertised as the primary behavioral health resource for service members in North Carolina and the gateway to other support services. Indiana reports a similar system of assisting service members through a single helpline and connecting them with the counselor assigned to their region, ensuring that clients are assigned and accounted for.

High-Touch, One-on-One Services

The majority of programs we assessed focus on providing high-touch, one-on-one services, and we believe this approach may be helpful for other programs, provided they have the resources necessary to accomplish it. For example, Vermont provides very personal outreach that includes transporting individuals to medical appointments and connecting individuals with a VA liaison who can help them navigate the VA system. In Washington, individuals are followed through every step of their job search; program staff provide support for preparing résumés, sharpening interview skills, and tailoring application materials to a specific opportunity, and then follow up with employers and human resource offices if possible to advocate for applicants and receive feedback on their qualifications.

Program staff in several states report that one-on-one, personal contact with individuals is critical to program success. Staff in several of the employment programs contend that technology-based efforts alone would not be successful in meeting the needs of service members and that the one-on-one services they provide are the key feature that distinguishes the programs from the existing landscape of job boards. The programs report that the feedback they have received from service members is that job boards are not viewed as helpful. Methods of communication vary widely, including telephone, email, social media messaging, and in-person meetings, and all of these communication efforts help to support the one-on-one relationships that counselors in many of the programs establish with their clients. Our findings are reinforced by the literature, which also indicates that high-touch, one-on-one services are a best practice for social services programs.[5]

---

5   Bloom, Hill, and Riccio, 2003; Nightingale et al., 2002.

Geographically Dispersed Staff

Placing program staff in various locations across the state rather than in one central location is another practice that we believe holds promise for other programs. In some states, program staff are assigned to specific geographic regions, and staff often work out of a location in that area. This simplifies counselors' efforts to connect with service members, employers, and community partners in the region and helps them to be more aware of region-specific needs. In addition, programs in Indiana, North Carolina, Tennessee, and Washington all reported that the decision to locate program staff in armories helps to support program success by affording them greater access to leadership and service members and increasing the visibility of the program.

Activities That Build Long-Term Client Skills

A promising practice we observed at several sites is an effort to provide services that not only meet immediate needs of service members but also help to build service members' skills so that they may be able to address future challenges independently, without the help of resource providers. For example, the Indiana and Washington programs both devote substantial efforts to providing career counseling, including translating military skills to civilian jobs, identifying potential areas of interest, moderating expectations about salary and job level, and encouraging service members to engage in networking activities to build careers. Program leadership in several states emphasized that what was important was to ensure that service members are eventually placed in "meaningful employment," or a "career rather than a job," even if initially the service members are placed in jobs that do not meet this threshold in order to overcome immediate financial issues and begin accumulating work experience. The program in New Hampshire similarly seeks to address both immediate needs and individual capacity to address future needs in areas such as suicide prevention, access to mental health care, employment, and homelessness. Similarly, the program in Vermont seeks to assist veterans in building life skills that will prevent them from experiencing recurring problems with issues such as unemployment and homelessness. Providing individuals with skills and training rather than simply addressing immediate needs has been shown in the literature to play a role in the success of other programs.[6]

Strong Partnerships with Other Resource Providers

Most of the programs report that partnerships with other resource providers contributed to program success. As an example of partnerships internal to the National Guard, the programs in North Carolina and Oregon report that they refer unemployed individuals to mental and behavioral health services programs when they identify potential mental health issues, and those programs reciprocate if their clients need employment

---

[6] Scott Hebert, Anne St. George, and Barbara Epstein, *Breaking Through: Overcoming Barriers to Family-Sustaining Employment*, Abt Associates Inc., February 2003.

assistance. These partnerships with other offices within the National Guard ensure that personnel receive the comprehensive services that they often need in a seamless way, even when the BYR program itself does not provide wraparound services. Partnerships with resource providers external to the National Guard similarly allow the BYR programs to link individuals to needed services that are not directly provided by the program. For example, New Hampshire reports that the Chaplain Emergency Relief Fund and the VA provide important support to CCP through supplementary services and referrals. The Tennessee program benefits from partnerships with the Tennessee VA, the Tri-Cities Military Affairs Council, and the Wounded Warrior Project. Vermont's veterans outreach program has extensive networks with providers throughout the state, allowing it to more effectively identify veterans in need.

Several states report that they make a concerted effort to eliminate duplication across programs by surveying the landscape of existing programs, designing the program to meet underserved needs, and developing partnerships with other organizations to create a support network for service members. For example, Indiana program staff are encouraged to outsource résumé development to Operation: Job Ready Vets, which already provides regular workshops for this service. Research on service program implementation highlights partnerships with other providers as an important best practice.[7]

Strong Partnerships with Community Organizations and Employers

In a related vein, establishing strong partnerships with community organizations and employers is a promising practice. For example, the North Carolina Legal Assistance program reported strong connections with the community, which has allowed it to recruit high-quality legal talent.

Partnerships with employers can play a role in the successful implementation of employment programs as well.[8] For instance, Washington's EEP has maintained connections with a large number of community partners, including labor unions and technical colleges, which have improved the program's ability to provide effective employment-related services. Businesses and schools in the community are also a source of support for EEP. The Indiana program values its partnerships with employers as being so important that it referred to having "two customers, one mission." According to staff in these programs, strong relationships with employers facilitate placement of personnel into jobs by helping to ensure that personnel are strongly considered for (or given preference for) the positions to which they apply. The Tennessee NGEEP assigns counselors to regions of the state, and they have considerable latitude both to develop relationships with employers and other organizations in their assigned region and to tailor their efforts based on these region-specific employer needs.

---

[7] Fischer, 2005; Nightingale et al., 2002.

[8] Fischer, 2005; Nightingale et al., 2002; R. Holm, "Aligning for Impact: The Milwaukee Area Workforce Funding Alliance," *Job for the Future*, 2013.

Some program staff expressed disappointment with prior efforts to identify all potentially interested employers and allow them to post positions; this activity resulted in some wasted time on the part of program staff and service members because some employers that expressed interest did not have a genuine commitment to hiring veterans. In California, program staff hold employers accountable by following up with them to understand why individual applicants were not hired, and employers who are viewed as lacking commitment to the program are dropped from counselor caseloads. Assigning employer caseloads is another signal of the California program's intense focus on building employer relationships. Counselors have ongoing relationships with specific employers; thus, counselors are familiar with the employers' needs and can better prepare candidates for interviews.

For related reasons, programs in several states have shifted away from job fairs to efforts that bring together a smaller number of employers and focus on deeper partnerships. These smaller events are viewed as successful because they include only employers that are the most committed to hiring service members and that are often ready to hire on the spot. The events also allow personnel to better prepare for on-site interviews for positions with the employers that participate.

Also of note, several of the employment programs reported that strong relationships with the human resource community have been useful in refining program activities and identifying potential placement opportunities. For example, Society for Human Resource Management members provided Washington EEP program staff with guidance on online résumés (e.g., what keywords and formatting to use).

Early Engagement with Mobilized Units

Another practice that could prove helpful to many programs is early outreach to units that are still mobilized. Staff from multiple programs reported that they were more likely to have success in achieving outcomes when their clients' needs were addressed early. To ensure that service members have a successful reintegration, many of the programs we assessed indicated that they first attempted to connect with units still in theater or within weeks of their return from deployment. For example, North Carolina's IBHS program reported that reaching out to guard personnel while still deployed is important so that personnel are aware of the program in case they need behavioral health support after returning home. Indiana's program also reaches out to units early in order to identify those individuals who may not have employment lined up for their return home. Studies of employment programs indicate that engaging with individuals early and placing them in jobs quickly can help program success.[9]

---

[9]   Bloom, Hill, and Riccio, 2003; Jessica Collura, "Best Practices for Youth Employment Programs: A Synthesis of Current Research," *What Works Wisconsin—Research to Practice Series*, No. 9, University of Wisconsin–Madison, August 2010.

Broad and Creative Outreach Strategies

Earlier, we identified free and low-cost technology and social media as facilitators of program success, and more generally, we consider the use of technology and social media to be a promising practice that other states' programs could adopt. Colorado's program, which is focused on marketing, has a particularly robust social media presence, including efforts on Facebook, LinkedIn, and Twitter, and uses Quick Response Codes to integrate its printed and Internet media. The program also uses marketing software and televisions in armories to disseminate information about state family programs. Several states use Facebook to market their programs, communicate with service members, create a community, and post jobs. LinkedIn is similarly used for promotion, and a few programs work with service members to build their own pages to support career development. Twitter appears to be a newer tool used by a few states, but it has not yet played a primary role in outreach. Several states report that they regularly monitor social media engagement (e.g., via Google Analytics), and in some cases, states have used or are planning to use these data to refine their social media efforts. Most of the states that use social media reported that it is an extremely important tool for connecting with personnel, particularly among younger populations, and those states that are not using social media reported that they intend to do more in this area.

In addition to using social media, programs report other outreach strategies that are important to informing individuals about the services offered. In-person relationships with command leadership and a visible presence in armories are important means of outreach for many programs, allowing counselors to connect directly with individuals. Program staff report that using a mix of communication strategies—including in-person meetings, telephone calls, texts, emails, and social media—is important to maintain visibility and ensure that the programs are able to reach all individuals needing services. Studies indicate that meeting individuals where they reside can be useful to engage them and encourage them to make use of a program's services.[10]

Strong Technology-Based Tracking Systems

Many of the programs we visited described using technology-based tracking systems to ensure that individuals were accounted for throughout the process. These systems are important because they help support the one-on-one nature of interaction with individuals, prevent anyone from slipping through the cracks, and enable ongoing measurement and evaluation of outcomes.[11] Some of the programs rely on internally created tools, such as spreadsheets, to track outcomes and the movement of individuals through the system. For example, several programs require counselors to update Excel

---

[10] Farhana Hossain, Peter Baird, and Rachel Pardoe, *Improving Employment Outcomes and Community Integration for Veterans with Disabilities*, New York: MDRC, 2013.

[11] Bloom, Hill, and Riccio, 2003; Fischer, 2005.

spreadsheets weekly to account for activities and outputs, and these individual spreadsheets feed into a central tracking spreadsheet.

Programs also rely on more-sophisticated technology-based tools. Several of the states with employment programs, including Florida, Indiana, and Oregon, use systems like H2H and NGEN as the primary database for tracking individuals, and program staff believe that the ability to leverage technology while continuing to provide in-person services is important. These databases were reported to be relatively user-friendly and capable of providing a range of benefits beyond tracking, such as connecting individuals to job opportunities and gathering information on potential employer partners. After initially operating with hard copy files, North Carolina's IBHS program currently uses VistA/CPRS as its case management system, which is the same system used by the VA; this has facilitated more-efficient case management. Overall, we regard the use of established technological systems like H2H, NGEN, and VistA/CPRS for tracking and case management as a promising practice that could be transferred to other programs.

Vermont's Veterans Outreach Program has also been a leader in developing metrics and strong tracking systems. One of the first staff members to be hired into the program was a program analyst, which was a key factor in developing data systems at the very outset of the program. This in turn has facilitated program improvement based on analyzing the data that the program collects. Other programs could benefit from employing a similar skill set in their core program staff.

Efforts to Make Programs Replicable and Sustainable

A few states have made concerted efforts to ensure the sustainability of their programs by carefully documenting program activities, including expectations for SOPs, effective or promising practices, challenges that the program has faced, and solutions identified to address these challenges. This is especially laudable given the lack of continuity within some programs, substantial autonomy left to individuals in other programs, and reports by programs that their success relies largely on individual personalities. Colorado, North Carolina, and Vermont are three states that have made efforts to create documentation to support the continuity of programs, a practice noted in the literature as one indicator of a high-quality program.[12] For instance, Colorado program staff created templates for the state family program staff to use and carefully documented processes. In North Carolina, IBHS established an SOP manual that is the backbone of the program. The manual is considered a living document and is continuously updated. In Vermont, VTVOP has established an SOP manual called "the Battle Book." All staff are trained on this book and use it as the main guide for the program's policies and procedures. Finally, the uncertain budgetary environment has caused the Vermont program to take additional steps to ensure program continuity. The program

---

[12] Acosta et al., 2014.

made a deliberate decision to prioritize program personnel and phase contracts so that the contracts for the most essential personnel will end last in case funding for the program is cut. These types of arrangements can help to sustain the program in times when funding is uncertain or delayed.

Areas for Improvement

In our analysis, we found several pressing areas for improvement that many of the programs can focus on. These include a lack of well-defined, measurable goals; insufficient evidence of outcomes and impacts; insufficient outreach to the entire eligible population; and no contingency plan for the limitations of BYR funding. We describe these areas for improvement in the sections that follow.

Lack of Well-Defined, Measurable Goals

Many programs in our study lack clear, measurable goals, and we regard this as a critical area for improvement. The literature identifies a clear mission and well-defined goals as important to success for programs that provide employment and mental health services,[13] and more generally, a program's goals should drive its activities and how it demonstrates results in both the short and long terms. However, we found that programs vary in the types of results they aimed to achieve. Some programs are focused on outputs rather than outcomes (e.g., providing legal services at low cost), some focus on outcomes (e.g., achieving a certain number of placements), and others focus on higher-level impacts (e.g., reducing unemployment rates). Several of the programs have established goals for which success cannot readily be measured (e.g., improving guard resilience and readiness), and other programs have little or no way to demonstrate progress in achieving their goals. Finally, some programs have goals that lack benchmarks—for instance, proposing to "provide services" or "achieve placements" without specifying the number of services or placements that would be indicative of program success.

In addition to setting clear, measurable goals, there is strong evidence in the literature that suggests the importance of clearly articulating how a program's goals will be achieved through a "logic model." This model outlines the activities the program will engage in and the links between these activities and immediate outputs, client-focused outcomes (short-, medium-, and long-term), and the higher-level impacts that the program has set out to achieve.[14] Many government offices, including the Department of

---

[13] Collura, 2010; Matthew Chinman, Pamela Imm, and Abraham Wandersman, *Getting to Outcomes™ 2004: Promoting Accountability Through Methods and Tools for Planning, Implementation, and Evaluation*, Santa Monica, Calif.: RAND Corporation, TR-101-CDC, 2004.

[14] Rossi, Lipsey, and Freeman, 2004; Knowlton, 2013; John A. McLaughlin and Gretchen B. Jordan, "Logic Models: A Tool for Telling Your Program's Performance Story," *Evaluation and Program Planning*, Vol. 22, No. 1, 1999.

Education and the Corporation for National Community Service, now require programs that they fund to develop a logic model prior to implementation. Logic models are useful because they help to ensure that activities are closely aligned with the goals of the program and can provide programs with a guide to the appropriate measures that can be tracked to assess program impact. Logic models can also be useful for program replication. We did not identify any program that had developed a logic model.

Insufficient Evidence of Outcomes and Impacts

Measuring outcomes and impacts is an important component of program evaluation that can be useful for external program validation and internal program improvement.[15] Many of the programs we assessed lack the data needed to determine whether they are meeting their goals. In some cases, this is because their goals are not well-defined and are therefore challenging to measure. In other cases, programs have set clear goals but are unable to collect the data needed to assess progress toward these goals. For example, Florida's program has set clear goals based on guard unemployment rates. These unemployment rates were estimated in the early years of the program based on monthly data that were pushed upward from individual units. However, in recent years, the program has not regularly collected these data. Programs in several other states also report that unemployment data are challenging to collect and are often unreliable because they do not account for voluntary unemployment (e.g., individuals enrolled in school).

In addition, most of the programs we examined collect little or no feedback from clients. Although such objective measures as job placements, tax returns prepared, suicide rates, and cost savings can be indicators that a program has met its goals, subjective data, such as the perceptions of program clients, can be useful as well. For example, understanding whether clients find their new employment to be rewarding and lucrative can help to assess underemployment. Among the few programs that do collect this information, the efforts reach a relatively small portion of the population served or are too new to provide evidence of program success and improvement. For example, program staff in New Hampshire indicated that they struggle to collect sufficient data and persuade participants to complete surveys. The North Carolina Legal Assistance program noted a similar problem with its satisfaction surveys. Programs could benefit from additional client feedback and should ensure that there are mechanisms in place to both obtain participant feedback and follow up to ensure high response rates. Ultimately, participants' perspectives and experiences with the program are an important element of gauging whether the program is successful. This information can also sup-

---

[15] Rossi, Lipsey, and Freeman, 2004; Harry P. Hatry, *Measuring Program Outcomes: A Practical Approach*, United Way of America, 1996.

port decisions on whether a program needs to take midcourse corrections, as well as which corrections are most appropriate.

However, simply tracking outcomes related to the program is not sufficient to measure program impact, even with high-quality data.[16] In order to measure impact, programs must be able to compare outcomes for program users with outcomes for a similar comparison group. The programs we assessed typically monitor outcomes only for those participating and do not have a comparison group. For example, many of the employment-focused programs focus on numbers of placements and absolute unemployment rates as indicators of success. However, without a comparison group, it is difficult to determine whether or when an individual would have found a job without the program or what the characteristics of the job would have been. It is also challenging to identify how much of a decline in unemployment rates can be attributed to the programs themselves as opposed to general economic trends. Several programs compare unemployment rates for the target population with unemployment rates for the states as one way of attempting to account for these economic trends. These benchmarks, while not sufficient to make causal statements about the impact of the program, provide a rough comparison group that can help to account for some of the factors other than the program that may be driving changes in outcomes.

Insufficient Outreach to the Entire Eligible Population

The BYR appropriations were intended to provide support to individuals in both the National Guard and the Reserves. Most of the programs we assessed reported that they do not turn away any service member or veteran, regardless of whether an individual has been deployed or what component they come from. In some cases (e.g., North Carolina's Legal Assistance program), the programs also support family members. However, several programs primarily target their services to guard members, and in at least one state, the program focuses disproportionately on the Army National Guard. At the same time, several programs noted that their ability to provide high-touch, one-on-one services to their existing target population is limited, so expansion may require additional resources or scaled-back services. For example, California is currently expanding services to the 60,000 reservists in the state and expressed concern about how to increase outreach to this new population that is five times as large as the guard population they have previously targeted.

Several programs also acknowledge that they have been less successful in serving individuals who are located in rural areas. According to staff from two of the employment programs, it can be more challenging to develop strong partnerships with

---

16 Rossi, Lipsey, and Freeman, 2004; Carolyn J. Heinrich, "Outcomes-Based Performance Management in the Public Sector: Implications for Government Accountability and Effectiveness," *Public Administration Review*, Vol. 62, No. 6, 2001; Lawrence L. Martin and Peter M. Kettner, *Measuring the Performance of Human Service Programs*, Vol. 71, Sage Publications, 1996.

employers in these regions that will result in placements, and as a result, building employer relationships is often focused on urban areas. The efforts of programs like Tennessee and Washington to assign geographic regions to counselors is one strategy that has been utilized to ensure sufficient attention is devoted to all regions of the state.

No Contingency Plan for the Limitations of BYR Funding

The majority of the programs in our study rely solely on BYR funding, and many expressed concerns related to short-term funding cycles and the year-to-year funding uncertainty. Funding issues have been a pervasive concern for many other programs that are funded by federal and state governments, which often rely on temporary grants and contracts, leading to instability and uncertainty.[17] However, while some programs had diversified funding to address uncertainty and supplement federal resources, most of the programs did not use or even search for funding from different sources to help sustain critical programs. Furthermore, even with these short-term funding cycles and uncertainty about whether additional years of funding would be available, some BYR programs spent a substantial portion of their initial year developing their programs rather than executing them fully and serving the full eligible population. In other words, the amount of time spent planning was potentially excessive relative to the time spent executing the program's mission, especially considering that the funding might soon evaporate. Further, even given the uncertainty of BYR funding continuity, some states used the appropriation to support, even fully, what may be regarded as essential or critical services.

---

[17] Nightingale et al., 2002; Eyster et al., 2010.

CHAPTER THIRTEEN

Conclusions and Recommendations

In this project, our objectives were to determine the extent to which the BYR programs are meeting their stated goals, identify transferrable promising practices, and suggest ways to improve program effectiveness. Our approach centered on a series of site visits, one to each program. We interviewed each program director and multiple staff members at each site. We also collected numerous documents detailing each program's goals, activities, outputs, data, policies, and other aspects as deemed relevant. The notes from these semistructured interviews were central to our research; the supporting documents added detail and granularity. We reviewed and integrated data from all of the sources, then developed program-specific assessments. Finally, we identified issues that cut across states; these informed our recommendations.

The vast majority of programs that we reviewed have at least partially met their stated goals. In one instance, we determined that a program had not met its stated goals; in another instance, we lacked sufficient information to determine whether the program met its stated goal. In most cases when programs failed to meet goals, this was at least partly due to the goals being unclear or immeasurable or because the program was very new. A lack of evidence that demonstrated progress toward a goal is also an issue for some programs.

Programs faced several challenges and exhibited some shortcomings, but we were also able to identify many transferrable promising practices. These should be helpful to programs as they strive to improve their effectiveness. In addition, based on our assessments of the 13 BYR-funded programs in our study, we have identified a series of recommendations for improving the effectiveness of these programs. Some of the recommendations apply to a subset of programs, but others are relevant for most or all programs. We also offer recommendations that may be helpful in assisting DoD or congressional policymakers as they consider future BYR funding allocations and general program oversight.

Program-Specific Recommendations

We have the following recommendations for program leaders:

- Develop meaningful, measurable goals.
- Collect and learn from program data on effectiveness.
- Ensure that programs are sustainable.
- Utilize practices associated with high-quality programs.

Develop Meaningful, Measurable Goals

Programs should first conduct a needs assessment, which would also include identifying gaps in the services available, defining their own purpose, and developing goals that are meaningful, measurable, and focused on concrete outcomes. As a second step, programs should ensure that they are collecting the relevant data to determine effectiveness. This recommendation is relevant for every program we assessed, to some extent.

Some programs have long-term goals focused on impacts, such as increasing readiness and resilience or decreasing the statewide unemployment rate among reserve-component personnel or veterans. While such goals are admirable, measuring progress toward such ambitious and wide-ranging goals is a challenge, primarily because of these programs' limited scope. For example, they may be able to affect their clients' readiness or resilience, but having a significant impact on total force resilience is less likely. Similarly, employment programs may yield positive employment outcomes for individual personnel, but they may not be able to influence major changes in statewide unemployment, particularly when there are other economic factors at play. Additionally, the link between the programs' activities and these goals is unclear, creating a situation in which some programs may never be able to achieve their goals. Therefore, we recommend that programs develop more-concrete goals and ways to collect information to measure progress toward those goals. Even programs with well-defined and measurable goals may benefit from including more-nuanced measures (such as general participant feedback or satisfaction levels) and measures that focus more on quality and less on quantity (such as the quality of jobs that service members find). When distinct measures, such as objective data and subjective information from clients, complement one another, that can give programs more confidence that their activities are yielding the desired outcomes than they would have by relying on one set of metrics alone.

Therefore, we recommend that the programs develop meaningful, measurable goals based on their specific activities and outcomes. Programs should seek to measure what they *do* (outputs) and what they *get* (outcomes). The programs we assessed offer some examples of concrete output metrics, such as website hits, presentations given, activities attended, telephone calls placed, résumé and interview skill sessions, referrals, counseling sessions per staff member, job applications submitted, and tax returns filed. The greater challenge is developing concrete outcome measures. We do not mean to imply that all programs should have the same goals or that all goals must be easily quantified (i.e., countable). Rather, we recommend that programs collect measures that provide a robust picture of outcomes, both short-term and longer-term.

Programs should also collect data on whether intermediate steps or outputs are met so that the programs can identify trends on steps toward meeting outcomes. For example, the bulk of the programs we assessed focused on employment. They can track short-term outcomes—such as the acquisition of skill certifications or occupational-specific credentials, as well as client perceptions of confidence in interviewing skills—as evidence of improving clients' employability. Further, employment programs could collect information not only on the *number* of placements, as many already do, but also on the placement's *quality* through such measures as hourly wage, the availability of benefits, and client satisfaction with the employment. In addition, while somewhat difficult, tracking clients' long-term employment outcomes, such as whether they still hold the job one year later, have been promoted, or have received a raise, may indicate that long-term outcomes have been achieved. While tracking these outcomes could be resource-intensive, it could yield much more helpful information than, for example, collecting unemployment data on service members. (Several of the employment-focused programs collect unemployment data, sometimes through the armories.) Looking beyond employment programs, keeping readiness and resilience in mind can be useful, but the focus and, consequently, measurement may be more on individual client-level outcomes rather than on overall force impacts.

Goals also need to be realistic, given both capacity and time frame (i.e., because annual funding is uncertain). For example, even if measuring readiness were a straightforward and low-cost activity, a small program is unlikely to affect the overall readiness of the state National Guard within a reasonable time frame. In some cases, developing a logic model or using another evidence-based method may be helpful as programs seek to align goals, activities, outputs, and outcomes. The first step in delineating a program model is to identify clear, measurable, and attainable goals. Once goals are determined, the program should identify a clear set of program activities, and these activities should be linked to the intended goals through a series of intermediate outcomes. Program effectiveness and efficiency can be improved when activities that are not aligned with program goals or cannot be linked to clear intermediate outcomes are eliminated, ensuring that the program is designed explicitly to meet the goals set forth and that resources are used in a focused and effective manner.

Collect and Learn from Program Data on Effectiveness

After setting appropriate, concrete, and measurable goals, the next logical step is to collect meaningful data. In this case, meaningful data are data that will allow programs to determine whether or not their goals are being met. Thus, data collection should flow directly from the program's goals. Collecting and reviewing such data should allow programs to pick up on trends and to make adjustments to resources as necessary.

Collecting such data, however, is not sufficient to determine causality; assessing whether these programs actually *bring about* desired changes requires another step. Linking the existence of programs to desired change requires developing an evaluation with a *control group*—a group of people who are very similar to program participants but who do not receive services. By tracking outcomes for program participants and for the control group, it is possible to determine the extent to which the program actually *causes* specific changes or outcomes. Such an evaluation generally would be carried out above the program level and by an outside group (rather than by program staff).

We do not mean to imply that programs fail to collect data. Indeed, many of the programs collect substantial amounts of data. However, we found that in some cases, their data collection efforts were either output-focused or not well aligned with their goals, and the data often were not used to make program improvements.[1] In short, many of the programs lack a culture of data collection, usage, and analysis. One way to encourage such a culture is for leadership to focus attention on data usage and analysis. Another strategy is to use established technological systems, which several programs already report has been successful and effective for tracking data. At least one program strategically hired a staff member with information technology and analytic expertise early in the program's life cycle.

A final point is that some programs face specific barriers to data collection. For example, they may lack the authority to collect the appropriate data, they may lack the time or resources to collect the data, or they may struggle with formatting or storage issues (such as data that exist only on paper or in difficult-to-aggregate formats). Program leadership or state-level interventions may be necessary to resolve these problems.

Ensure That Programs Are Sustainable

The uncertain nature of BYR funding poses a substantial challenge for all of the programs; a loss or delay of funding would likely mean at least a temporary closure of many programs and thus the loss of program services. On a related note, some of the BYR programs are small, and substantial amounts of program-specific knowledge reside with individual employees; employee turnover in such programs could mean the loss of significant levels of knowledge and of relationships with stakeholders. This is especially likely to be the case when programs allow staff significant autonomy in their jobs. In many cases, such autonomy is appropriate or even a promising practice.

While program directors have only limited ability to control funding or turnover, there are practices that can serve to make programs more sustainable even in light of such uncertainty. For instance, programs may be able to establish alternative funding streams or sources, which could serve as buffers against funding uncertainty. Creating and using SOPs can also help to retain knowledge even in the face of staff turnover.

---

[1]   Some programs do use some data elements to track progress, and most programs do make use of some data; we explicitly discuss each case in the individual program chapters.

Programs will likely benefit from clearly defined staff roles, and some programs may benefit from providing additional staff training. Finally, encouraging or requiring program staff to keep detailed records of contacts with service members and other stakeholders may help to improve program sustainability from a case management standpoint, even in the face of staff turnover.

Utilize Practices Associated with High-Quality Programs

We recommend that programs utilize practices associated with high-quality programs. In the literature, high-quality programs are described as ones that are evaluated, sustained, and replicable. Accordingly, the aforementioned use of SOPs is associated with high-quality programs because these help to ensure the program will persist in light of staff turnover and that the same successful practices will be used consistently. Further high-quality program practices include making efforts to ensure program viability even in light of BYR funding decreases and ensuring that services provided to clients are evidence-based. Finally, using an evaluation process (with a control group, as described) is linked to improved decisionmaking, oversight, and monitoring.

Recommendations for Department of Defense and Congressional Policymakers

We have the following recommendations for DoD and congressional policymakers:

- Address programs' concerns regarding BYR funding.
- Clarify appropriate use of BYR funding.
- Share promising practices across programs.
- Encourage programs to widen their focus beyond the National Guard.

Address Programs' Concerns Regarding BYR Funding

The uncertainty of BYR funding stems in part from a fluctuating federal budget environment. However, this uncertainty may negatively influence several outcomes, from long-term planning to potential commitment of staff members. Indeed, in some cases, funding issues have caused programs to delay or stop and then restart operations. Therefore, we recommend that DoD and congressional leaders do what they can to ensure that, as much as possible, funds are transferred in a timely manner. Also, leadership should consider ways to encourage, or require, programs to plan for continuity of operations in cases of uncertain funding (clear rules and SOPs may be especially helpful). This high-quality program practice is especially important in the BYR funding environment.

Clarify Appropriate Use of BYR Funds

The uncertainty of BYR funding also suggests that it may be especially appropriate to think strategically about the types of programs BYR funds support, and, more generally, about how BYR funds should be used. Some programs funded with BYR dollars may be viewed as critical while others may be considered supplemental. For a critical program, the end of services or even a lapse in services could be extremely problematic. For example, when suspending services to a BYR program that is the only resource in the state to provide a critical service (e.g., suicide prevention), the losses from lapsed service may be difficult to reverse.

Along the same lines, DoD and congressional leadership should consider what are the most appropriate uses of these BYR funds. For example, the funds may work well for pilot programs that intend to find alternative sources of funding if they prove successful. BYR funds also may be suitable for purchasing equipment to support service members with reintegration or to augment existing programs (rather than hiring temporary staff or standing up new programs). These options are intended to convey the range of ways funds could be used, all of which conceivably could support the intent of easing service members' transition to civilian life. Ultimately, DoD and congressional policymakers should consider providing guidance to the states on more—and less—appropriate uses of BYR funding.

Share Promising Practices Across Programs

We found that programs are very eager to learn what other programs are doing. Therefore, sharing information on promising practices, as well as barriers and the extent to which different programs have overcome barriers, is likely to be of interest to all programs and could be especially helpful for new programs. The BYR-funded programs provide different services to different populations; therefore, promising practices will not be "one size fits all." However, many of the programs have developed some innovative, promising practices. Examples include close alignment between goals and activities, a single point of entry, strong relationships with a wide variety of stakeholders, high-touch services, early engagement, and broad outreach strategies (discussed in more detail in Chapter Twelve). We encourage DoD and congressional policymakers to share this information in a variety of ways, such as encouraging cross-program communication and collaboration (perhaps via web-based meetings or an in-person meeting for program staff and leaders). Sharing barriers, even if programs have not found ways to overcome these issues, may be useful as well.

Encourage Programs to Widen Their Focus Beyond the National Guard

Some programs focus primarily on National Guard, or even Army National Guard, personnel in their states. This is not surprising, because the appropriations generally are distributed to the individual state National Guard units, and the programs are often located within National Guard commands. Programs largely report that other service members, veterans, and, in some instances, family members were included in their target populations, and some programs report making a concerted effort to serve their entire target population. However, in many cases, the programs are National Guard–centric or Army National Guard–centric. We recommend that leadership encourage programs to widen their focus and their marketing efforts to include more service members, veterans, or family members as appropriate. This may be particularly suitable in states that have dispersed or remote populations, and may include providing more-dedicated outreach and services to individuals in rural and other underserved areas. If it is the intention that these programs serve both the National Guard and Reserves, this should be communicated to programs that are currently focused primarily on the Guard, and efforts should be made by the programs to expand services. We note that some programs, like California's Work for Warriors, are making efforts to expand their services to include their entire target population.

A similar issue involves many of the programs that focus on employment. The majority of BYR employment-focused programs report collecting some form of unemployment data on National Guard personnel, but no program explicitly mentioned USERRA, and only one program mentioned a partnership with ESGR. USERRA generally ensures reemployment at a service member's former civilian job upon the completion of deployment service, and ESGR can provide assistance to service members in retaining their prior positions. Thus, if they are not already doing so, employment programs may find it valuable to refer some service members to ESGR for assistance in retaining their prior jobs. Also, to the extent that programs wish to continue to collect unemployment data, including questions about the service member's eligibility to return to his or her former job and reasons why a service member may not wish to return to that job could provide valuable information to help programs more carefully tailor their services.

Study Approach

The programs included in our study were selected by our research sponsor and represent the full set of states that received BYR funding in FY13. We used a standardized case study approach for our assessment. Although the generalizability of our results is limited, this research design enabled us to accommodate for differences in individual program scope and focus while still providing the basis for a systematic, rigorous program assessment.[1] We collected comparable information about each program, integrated different data sources to assess the program, and then looked across the programs for cross-cutting themes, such as common facilitators and areas for improvement.

Each program case study consisted of the following elements, with a site visit to each program headquarters central to our approach:

- pre-site visit interview with the program point of contact
- review of available program documentation, such as a program logic model or conceptual framework, marketing materials for prospective clients, records of program outputs or outcomes, and program evaluation or audit reports
- on-site interviews with program staff
- on-site feedback session with program staff.

The site visits were made possible through the efforts of the National Guard Bureau. Specifically, BG Marianne Watson (director, Manpower and Personnel, National Guard Bureau) sent a letter to the adjutants general in all the states included in our study. In her letter, Brigadier General Watson explained the purpose of the study and requested that the states provide RAND with a point of contact with whom the RAND study team should coordinate their site visit and data collection requests.

---

[1] The study was submitted for review to RAND's Institutional Review Board, also known as RAND's Human Subjects Protection Committee. On November 26, 2013, the committee determined that the study did not involve human subject research. On December 10, 2013, the Office of the Under Secretary of Defense for Personnel and Readiness concurred with that determination.

To develop the interview protocols for this project, we consulted with literature related to program review, program evaluation, and logic models.[2] The protocols (provided at the end of this appendix) covered the following topics:

- program background: such details as how long the program has been in place and how it is administered
- goals: what the program aims to accomplish
- inputs: resources, such as people, money, and materials
- activities: what the program does to reach and help guard and reserve personnel, veterans, and other clients
- outcomes: how clients are affected by the program
- impacts: how the state, DoD, and society are affected by the program
- facilitators and challenges: what has helped the program to meet its goals, and what has made it harder to do so
- promising practices: practices that have worked well for the program and may be useful to other programs.

We began our data collection with a brief telephone interview prior to each site visit. We obtained basic information about the program's goals, activities, and available documentation that could help us learn more about the program in advance of our visit. Two RAND researchers visited each site, typically at the state's National Guard Joint Force Headquarters, and met with a program's director and other staff. Across programs, the number of people we met with ranged from two to nine people. While on site, one RAND researcher conducted the interview and the other took detailed notes. Our general approach was to cover questions about program history, goals, resources, activities, and results in an initial interview session. Following that, we took a break to rapidly prepare a set of slides that summarized what we learned during that session. Teams used a preestablished briefing template to develop the slides, which loosely followed the steps one would take to develop a program logic model, identifying what the program wanted to do (goals) and actually did (activities), what resources it relied on to achieve its goals, how it measured usage (output), and how it measured success (outcomes and impact). When we reconvened for our second interview session, we presented the slides to the program director and staff, asking for corrections and using them to elicit more information about the program. Finally, in a typical visit, we wrapped up the site visit with questions about data collection and usage, program facilitators and barriers, and future plans. There were some exceptions to this approach—at times, the order of topics covered varied, and sometimes the feedback session occurred at the end of the visit instead of the middle—but for all programs, we covered the full

---

[2] The works we relied on the most were Acosta et al., 2014; Knowlton and Phillips, 2013; and Rossi, Freeman, and Lipsey, 1998.

set of interview questions and conducted a feedback session. Representatives at every site visit also had an opportunity to comment on our preliminary impressions.

Soon after each site visit, we finalized interview notes, which set the stage for our within-program analysis. Following Eisenhardt's work,[3] our within-program analysis centered on detailed program write-ups intended to be primarily descriptive, and it did not include researcher impressions and other commentary. To ensure the program write-ups had a parallel structure that would facilitate cross-program analysis, we developed a program synthesis guide. This guide, which is included at the end of this appendix, provided detailed instructions about the program write-up process and a list of topics that the write-up should cover. To promote external validity for this qualitative analysis,[4] the two RAND researchers who participated in each site visit independently reviewed the interview notes and documentation for that program and then drafted their analysis of patterns and cogent findings for the program. The program synthesis guide topics corresponded to those featured in the interview protocols (e.g., program activities, outputs, goals, outcomes, facilitators, challenges) and also covered background and orientation information, areas for improvement, and promising practices. The authors then reviewed one another's work and consolidated their analyses into one document, resolving conflicting data points through discussion and subsequent revisions. Overall, the independent write-ups were complementary; few contradictory findings or conclusions were noted. The consolidated write-up was then forwarded to a third RAND researcher for review. After all third-party reviews and resultant revisions were completed, the program write-ups were expanded to develop the detailed assessment for each state's programs (Chapters Two through Eleven).

The program write-ups also served as the basis for the cross-program analysis. Specifically, we "coded" the write-ups based on the topics in the program synthesis guide. As part of a formal qualitative analysis, codes are used to retrieve and organize data by topic and other characteristics. After the coding was complete, pairs of RAND researchers again independently reviewed the results and drafted common themes and notable findings. For example, two researchers reviewed all the sections coded as "areas for improvement" and noted those that were common across several programs. The results of this cross-program analysis for facilitators, challenges, promising practices, and areas for improvement served as the basis for Chapter Twelve.

---

[3] Kathleen M. Eisenhardt, "Building Theories from Case Study Research," *Academy of Management Review*, Vol. 14, No. 4, 1989.

[4] R. B. Johnson, "Examining the Validity Structure of Qualitative Research," *Education*, Vol. 118, No. 2, 1997.

BYR Program Telephone Interview Protocol (Before Site Visit)

1. When did [PROGRAM NAME] start?
   a. Probe: Did it operate under another name before, or as part of a larger program? If yes, please tell me about how [PROGRAM NAME] evolved to its present state.
2. What are [PROGRAM NAME]'s goals and objectives?
3. What activities or services constitute the program?
4. How will those activities achieve [PROGRAM NAME]'s goals?
5. Do you have anything that describes how the different program activities are expected to lead to different impacts? Do you have any documents you could send me that illustrate this?
   a. For instance, do you have a process map or conceptual framework depicting how what the program does yields the intended results?
6. Does [PROGRAM NAME] have a website?
   a. [If affirmative:] What is the URL?
7. What types of documentation accompany the program, such as program manuals and brochures?
   a. [If documentation exists:] Can you provide copies of them to us?
8. What data or records are maintained for [PROGRAM NAME]?
   a. Prompt if needed: For example, do you track costs? Services delivered? Service quality? Outcomes? Something else?
   b. [If affirmative:] Are there reports or other records that you can share with us?
9. Has [PROGRAM NAME] been evaluated before?
   a. [If affirmative:] Are there reports or other records from that evaluation/those evaluations that you can share with us?
10. As part of RAND's study, we plan to make one site visit to conduct interviews in person and to learn more about the context in which [PROGRAM NAME] operates.
    a. Over the next few months, what timing works especially well—or especially poorly—for you and your colleagues?
    b. Aside from you, who else would you recommend I speak with to get a complete sense of your program?
11. May I answer any questions for you at this point?

BYR Program Site Visit Interview Protocols

Site Visit Session 1

1. To start, please tell us your job title and main responsibilities at [PROGRAM NAME].
2. What is [PROGRAM NAME] trying to accomplish? What are its mission and goals?
   a. What impact does [PROGRAM NAME] seek to have at the state level?
3. We'd like to learn more about [PROGRAM NAME]'s target population. Would you please tell me who is part of the target population, and how large that population is?
   a. Prompt if needed: An estimate of the population size is sufficient.
4. How many [target population] does [PROGRAM NAME] serve? [If not obvious:] Is that measured per year, month, week, day?
5. How broadly is [PROGRAM NAME] currently being implemented in [STATE]?
   a. Prompt if needed: For example, how many locations does the program have? Are all [target population] in [STATE] eligible?
6. Thank you for that information. Our next questions are intended to help us understand the resources that help [PROGRAM NAME] achieve its goals. First, how many paid staff members do you have working in the program?
   a. How many are full time and part time?
   b. What is their background or educational training?
   c. How long have the staff members worked with this program?
7. How many volunteers support [PROGRAM NAME]?
   a. What is their background or educational training?
8. What is [PROGRAM NAME]'s total annual budget?
   a. Prompt if needed: We're interested in current year as well as any future budget projections that are available.
9. How has [PROGRAM NAME]'s annual budget changed over the years?
   a. Prompt if needed: We'd like to understand how funding has varied over the program's life span.
10. How is [PROGRAM NAME] funded?
    a. Prompt if needed: What sources of funding does [PROGRAM NAME] have in addition to the funding from Congress, if any?
    b. [For each type of funding:] Is this one-time or recurring funding?
    c. [For each type of recurring funding:] How long do you anticipate the funding will last?

11. Does [PROGRAM NAME] receive any in-kind donations?
    a. Prompt if needed: Examples of in-kind donations include office space, donated advertising, pro-bono financial or legal services, website design or hosting, and personnel paid for by other organizations.
    b. [If affirmative:] What kinds, and from whom?
12. How does [PROGRAM NAME] raise funds and obtain those donations?
    a. Prompt if needed: For example, does [PROGRAM NAME] apply for grants? Solicit donations from specific organizations or people? Hold fund-raising drives?
13. Are there other resources that [PROGRAM NAME] relies on that we haven't discussed?
    a. [If affirmative:] Please tell me about them.

Program Activities

14. Our next set of questions is intended to help us understand what [PROGRAM NAME] does to achieve the goals you mentioned earlier. First, what activities or services constitute [PROGRAM NAME]?
    a. [For each activity/service ask (if cannot be inferred)]:
        i. Why is this activity/service used?
        ii. What is the mode of delivery? For example, are services provided face-to-face or over the telephone? Individual or group basis?
        iii. What is the frequency or intensity of the program? For instance, how many sessions does a client participate in? How long are they? How often do they occur?
        iv. Are veterans the sole target audience, or do other types of people participate as well?
15. [If not answered in Q14:] What types of problems does [PROGRAM NAME] address most frequently in these activities and services?
16. How does your target population learn about these activities and services? What kinds of outreach does [PROGRAM NAME] do?
    a. Do you attend any military-affiliated events to advertise [PROGRAM NAME]'s services?
17. Before we adjourn, we'd like to ask a few more questions about program activities. What happens as a result of [PROGRAM NAME]'s activities?
    a. Prompt if needed: How do they relate to [PROGRAM NAME]'s mission or goals?
18. What final outcomes are [PROGRAM NAME] trying to impact?
19. Does [PROGRAM NAME] expect some outcomes to be achieved immediately, and others to emerge over time?
    a. Please explain. [If affirmative:] Which outcomes are achieved in the short term and which take longer to realize?

20. What results have been produced to date?
21. What results are expected in the next two to three years?
22. Have any of [PROGRAM NAME]'s activities or strategies been particularly successful? Why?
23. We have just one more question for you before we break. Earlier you mentioned that [PROGRAM NAME] [refer to Q2a response about intended state-level impact]. What difference has [PROGRAM NAME] made at the state level?

Site Visit Session 2
Program Impact

24. During our break, we reviewed the responses you provided to our questions and created a process model that we believe illustrates what [PROGRAM NAME] does and the results that it seeks to obtain. To kick off our discussion [this morning/this afternoon], we'd like to walk through our model with you and make any changes you deem necessary.
    [INTERVIEWER: Walk interviewee through process map, working through resources, activities, outputs, outcomes, and impact.]
25. Thank you for ensuring that the process model we created based on your earlier responses is accurate. In our next few questions, we'd like to focus on how [PROGRAM NAME] knows how well it's doing in meeting its goals. To start, what evidence is necessary to determine whether [PROGRAM NAME]'s goals are met?
26. Has [PROGRAM NAME] collected that evidence? What types of information does [PROGRAM NAME] collect to determine if it is meeting [specify goals]?
    a. Prompt: This could include [if program has a website:] website activity, number of people that participate in the program, number of placements, indicators of client satisfaction, or cost savings. [INTERVIEWER: Include or substitute examples relevant to the specific program as available, with emphasis on outcomes.]
    b. [If affirmative:] In what manner are these types of information collected? [INTERVIEWER: Probe to obtain manner for each type.]
    c. [If affirmative:] How often are these types of information collected? [INTERVIEWER: Probe to obtain timing for each type.]
27. [If Q26 responses are primarily indicators of usage or access, and not outcomes:] You've given us some examples of information related to usage [or access]. How does [PROGRAM NAME] measure the ultimate impact of its activities?
    a. Prompt if needed: Does [PROGRAM NAME] conduct any sort of cost-benefit analysis or look at return on its investment. If yes, please tell us more about that.

28. How is this information used?
    a.   Prompt if needed: Does [PROGRAM NAME] summarize or analyze this information in any way? If yes, how?
29. Has [PROGRAM NAME] ever been the focus of an evaluation?
    a.   [If affirmative:] Please tell us more about that evaluation/those evaluations. For example, who conducted the evaluation, what types of information were collected, and how were the results used? [INTERVIEWER: We asked this in the initial telephone conversation; if you are only talking to the same person, don't ask—although this may be a good time to ask for evaluation results if they exist and we don't have them.]
30. As a result of [PROGRAM NAME]'s own analysis or any outside evaluations, were any areas for improvement identified?
    a.   [If affirmative:] Please tell me about them, and what [PROGRAM NAME] did or is doing to address them.
31. On the flip side, were any success stories or promising practices identified?
32. [If affirmative:] Please tell me about them, including whether [PROGRAM NAME] made any changes to promote their use.
33. Has anything changed based on the information [PROGRAM NAME] has collected [if applicable:] and on the results of the analysis and evaluation conducted about [PROGRAM NAME]?
34. What kinds of program performance information have been requested by the program's key stakeholders?
35. We've touched on this already, but I would like to ask explicitly, what does [PROGRAM NAME] plan to do if its goals are not met?

Facilitators and Challenges

36. Thank you for your time. We have just a few more questions for you. What factors help [PROGRAM NAME] in providing support to its target audience or achieving its goals?
    a.   Prompt if needed: This may include, for example, state-level support, trust within the community, and effective marketing or advertising.
37. What challenges has [PROGRAM NAME] faced in providing support to its target audience or otherwise achieving its goals?
    a.   Prompt if needed: This may include, for example, lack of knowledge about the program among target population, insufficient staff, or funding uncertainty.
    b.   What has or is [PROGRAM NAME] doing to overcome those challenges?

Future of the Program

38. Thanks again for your time and all the information you provided today. In closing, we'd like to discuss briefly the future of [PROGRAM NAME]. We know that [PROGRAM NAME] started in [year of inception]. Do you plan to make any changes in the program for purposes of sustainment?

39. What plans, if any, does [PROGRAM NAME] have to grow? This could include offering more activities and services, adding locations, or reaching a larger target audience.

40. In closing, is there anything else you would like to tell me about [PROGRAM NAME] that we haven't discussed?

Program Synthesis Guide

The purpose of this guide is to provide the backbone for a systematic examination of the evidence collected about each program. For each substantive heading or topic, team members should independently evaluate the data collected and craft roughly five to seven bullets (sometimes more, sometimes less) that capture the most compelling themes for that topic. The goal is to identify themes within and across data sources, primarily from interviews but also, to the extent possible, from materials provided by program staff and external sources. This summary should include analysis and assessment (i.e., use of expert judgment) but should also use references to evidence in support of such conclusions. The results of this exercise will serve as the foundation for the program's chapter in the final report, as well as any cross-program analysis we opt to do for a concluding chapter.

Special attention should be paid to identifying findings with a high level of agreement across data sources, as well as to those with striking contrasts between data sources (e.g., conflicting perspectives provided by interview subjects). The absence of data or a consistent lack of a response to a question may also be worth noting. Less emphasis should be placed on findings that emerge from only one data source; such findings are likely more tentative and possibly best presented as useful orienting background or interesting or unique features, where they will appear more as analytic "food for thought." Also, please clearly note areas in which opinions or assessments are based on statements by the program personnel versus those assessments based on team members' analyses.

While not all bullets need annotation in terms of sources, it likely will be helpful to note how prevalent a finding was (e.g., if multiple sources suggested it) or to cite a specific interview in the event of a controversial or unique point that others may wish to look up directly. This also helps to ensure that findings are selected based on evidence quality rather than salience or another bias.

Data Sources

Provide a short list of the sources that contributed to this summary, including a breakdown of interview subjects [including title/role], website URL [if applicable], and any documents from or about the program. Any gaps in data collection or data availability should also be noted here.

1. Useful Orienting Background

Detail any interesting, potentially relevant background that may help set the stage for subsequent findings. Include how broadly the program is offered within the state and where its staff are located. May include the history of the program, how it relates to other state programs intended to support guard and reserve personnel, how it came to receive BYR funding, and any plans to contract or expand.

2. Resources

List the resources available to support the project's goals and any information about how they have varied over time. Financial information should include not only budget but also funding sources and how funding is obtained. Human capital information should cover number, experience, and expertise of paid staff and volunteers, as available. Also include additional resources, such as donated goods and services. Finally, discuss any concerns about resource stability or plans to increase resources.

3. Activities and Outputs

Describe the program's activities. Include details such as the purpose of each activity (e.g., what need met, what problem solved or avoided), what output it produces, how it is delivered, and at what intensity. Also include any changes to the activities offered (e.g., different mode of delivery) and future plans for changes to existing activities or new activities. Fundraising activities should be covered under Resources, and outreach efforts go in Target Population and Outreach.

4. Target Population and Outreach

This section should cover details about the target population: type (guard/reserve, veteran, civilian, dependent), number served (actual numbers and as a percentage of overall population), any changes over time, and any plans to increase the number or type of clients/users. Also include how the program is marketed to prospective clients/users, noting in particular any tie-ins to YRRP or military-affiliated events.

5. Goals, Impacts, and Outcomes

Describe the program's mission and/or goals; how the program's activities relate to those goals; what outcomes and impacts the activities are intended to affect, and at what point in time (e.g., short term, long term); what state-level impacts the program seeks to achieve; results produced to date; and projected or expected future results.

6. Measurement and Evaluation—Information Collection

Describe how the program determines whether it has met its stated goals, including the evidence needed to do so. Also list and briefly describe the types of indicators that are collected for and/or by the program, noting in particular how they are collected, how often they are collected, and if they are usage-, access-, or outcome-related. Include relevant information from any external evaluations that have been conducted.

7. Measurement and Evaluation—Information Usage

Discuss how indicators of program usage, access, and/or impact are used by or for the program (e.g., analyzed for patterns or trends, cited as a basis for change, shared with program stakeholders). Note expressly whether any sort of cost-benefit or return-on-investment analysis was conducted. Include relevant information from any external evaluations that have been conducted.

8. Facilitators and Challenges

List and briefly describe key facilitators and challenges related to the program's successful implementation, as well as any successful strategies used to avoid or overcome challenges.

9. Areas for Improvement and Best Practices

List and briefly describe areas for improvement and best practices, including those identified during any evaluations or self-assessment. Examples of activities or programs that have been particularly successful should be featured here as well.

10. Program Interesting or Unique Features

Capture unique or noteworthy aspects of this program, implications of this case for cross-program analysis and for the overall study, and interesting observations that may be thought-provoking yet not strongly supported by the data sources.

References

Acosta, Joie, Gabriella C. Gonzalez, Emily M. Gillen, Jeffrey Garnett, Carrie M. Farmer, and Robin M. Weinick, *The Development and Application of the RAND Program Classification Tool, The RAND Toolkit*, Volume 1, Santa Monica, Calif.: RAND Corporation, RR-487/1-OSD, 2014. As of February 18, 2015:
http://www.rand.org/pubs/research_reports/RR487z1.html

Baldwin, David S., Adjutant General, California National Guard, "Employment Initiative for California National Guard Members," memorandum, December 27, 2013.

Becker, D. R., S. R. Baker, L. Carlson, L. Flint, R. Howell, S. Lindsay, M. Moore, S. Reeder, and R. E. Drake, "Critical Strategies for Implementing Supported Employment," *Journal of Vocational Rehabilitation*, Vol. 27, No. 1, 2007.

Bloom, Howard S., Carolyn J. Hill, and James A. Riccio, "Linking Program Implementation and Effectiveness: Lessons from a Pooled Sample of Welfare-to-Work Experiments," *Journal of Policy Analysis and Management*, Vol. 22, No. 4, 2003.

Chinman, Matthew, Pamela Imm, and Abraham Wandersman, *Getting to Outcomes™ 2004: Promoting Accountability Through Methods and Tools for Planning, Implementation, and Evaluation*, Santa Monica, Calif.: RAND Corporation, TR-101-CDC, 2004. As of February 18, 2015:
http://www.rand.org/pubs/technical_reports/TR101.html

Collura, Jessica, "Best Practices for Youth Employment Programs: A Synthesis of Current Research," *What Works Wisconsin—Research to Practice Series*, No. 9, University of Wisconsin–Madison, August 2010.

Congressman Mark Takano, "Rep. Mark Takano and Rep. Paul Cook Form Bipartisan Work for Warriors Caucus," Riverside, Calif., November 8, 2013. As of February 12, 2015:
http://takano.house.gov/media-center/press-releases/
rep-mark-takano-and-rep-paul-cook-form-bipartisan-work-for-warriors

———, "U.S. House of Representatives Passes Two Pieces of Legislation Submitted by Rep. Mark Takano," Riverside, Calif., May 22, 2014. As of February 12, 2015:
http://takano.house.gov/media-center/press-releases/
us-house-of-representatives-passes-two-pieces-of-legislation-submitted

Eisenhardt, Kathleen M., "Building Theories from Case Study Research," *Academy of Management Review*, Vol. 14, No. 4, 1989, pp. 532–550.

Eyster, Lauren, Demetra Smith Nightingale, Burt S. Barnow, Carolyn T. O'Brien, John Trutko, and Daniel Kuehn, *Implementation and Early Training Outcomes of the High Growth Job Training Initiative: Final Report*, Washington, D.C.: Urban Institute, 2010.

Family Program Marketing Office, *FY 2014 Standard Operating Procedure and Marketing Handbook*, draft February 2014.

Fischer, David Jason, *The Road to Good Employment Retention: Three Successful Programs from the Jobs Initiative*, Baltimore, Md.: Annie E. Casey Foundation, 2005.

Hatry, Harry P., "Measuring Program Outcomes: A Practical Approach," United Way of America, 1996.

Hebert, Scott, Anne St. George, and Barbara Epstein, *Breaking Through: Overcoming Barriers to Family-Sustaining Employment*, Abt Associates Inc., February 2003.

Heinrich, Carolyn J., "Outcomes-Based Performance Management in the Public Sector: Implications for Government Accountability and Effectiveness," *Public Administration Review*, Vol. 62, No. 6, 2002, pp. 712–725.

Holm, R., "Aligning for Impact: The Milwaukee Area Workforce Funding Alliance," *Job for the Future*, 2013.

Hossain, Farhana, Peter Baird, and Rachel Pardoe, *Improving Employment Outcomes and Community Integration for Veterans with Disabilities*, New York: MDRC, 2013.

Indiana National Guard, "Fact Sheet," web page, undated. As of October 15, 2014:
http://www.in.ng.mil/Portals/0/PageContents/ForMedia/INNGFactSheet2014.pdf

Johnson, R. B., "Examining the Validity Structure of Qualitative Research," *Education*, Vol. 118, No. 2, 1997, pp. 282–292.

Knowlton, L. W., and C. C. Phillips, *The Logic Model Guidebook: Better Strategies for Great Results*, Thousand Oaks, Calif.: Sage Publications, 2013.

Martin, Lawrence L., and Peter M. Kettner, *Measuring the Performance of Human Service Programs*, Vol. 71, Sage Publications, 1996.

McLaughlin, John A., and Gretchen B. Jordan, "Logic Models: A Tool for Telling Your Program's Performance Story," *Evaluation and Program Planning*, Vol. 22, No. 1, 1999, pp. 65–72.

Military Leadership Diversity Commission, *From Representation to Inclusion: Diversity Leadership for the 21st-Century Military*, Washington, D.C., March 2011.

Nightingale, Demetra Smith, Nancy M. Pindus, John Trutko, and Michael Egner, *The Implementation of the Welfare-to-Work Grants Program*, Washington, D.C.: Urban Institute, 2002.

Pogoda, Terri K., Irene E. Cramer, Robert A. Rosenheck, and Sandra G. Resnick, "Qualitative Analysis of Barriers to Implementation of Supported Employment in the Department of Veterans Affairs," *Psychiatric Services*, Vol. 62, No. 11, 2011.

Rossi, Peter H., Mark W. Lipsey, and Howard E. Freeman, *Evaluation: A Systematic Approach*, Thousand Oaks, Calif.: Sage Publications, 2004.

Ruzek, J. I., B. E. Karlin, and A. Zeiss, "Implementation of Evidence-Based Psychological Treatments in the Veterans Health Administration," in R. Kathryn McHugh and David H. Barlow, eds., *Dissemination and Implementation of Evidence-Based Psychological Interventions*, Oxford: Oxford University Press, 2012.

Sanders, Bernie, "Vermont Veterans Outreach Program Wins Funding," United States Senator for Vermont website, May 13, 2013. As of February 17, 2015:
http://www.sanders.senate.gov/newsroom/press-releases/
vermont-veterans-outreach-program-wins-funding

Scott, Rick, "Employment Initiatives Bring More Than 400 Jobs to Florida Guardsmen," Florida Governor Rick Scott website, 2012. As of February 17, 2015:
http://www.flgov.com/employment-initiatives-bring-more-than-400-jobs-to-florida-guardsmen-2/

Shepard, Ty, director, California National Guard Employment Initiative, "Lowering the Rate of Unemployment for the National Guard and Reserve: Are We Making Progress?" testimony before a hearing of the House Committee on Veterans Affairs, March 14, 2013.

Tyre, James D., Assistant Adjutant General, Florida Army National Guard, "Putting America's Veterans Back to Work," testimony before the House Committee on Veterans Affairs, June 2011.

U.S. Department of Labor, *Veterans' Employment and Training Services Request for Funding—Tab D: Enhanced Employment Program Return on Investment*, February 29, 2012.

Vermont Veterans Outreach Program, "Vermont Veterans Outreach Program Brief," 2013.

Yellow Ribbon Reintegration Program, *Annual Advisory Board Report to Congress, Fiscal Year 2011*, Washington, D.C., March 2012a.

———, Supplemental Funding Update briefing, August 20, 2012b.

YRRP—*See* Yellow Ribbon Reintegration Program.